QUANTUM SPIRITUALITY

"*Quantum Spirituality* stirs me in a special way. I love Peter's concepts relating to the mysterious and mystical aspects of life. It resonates with my own experience. It is a pleasure for me to see the remarkably incisive knowledge that Peter has crystallized in *Quantum Spirituality* to help us see and accept the true underlying nature of life. Please read it, and let it become a part of your life experience too."

BERNIE SIEGEL, M.D., AUTHOR OF *NO ENDINGS, ONLY BEGINNINGS* AND *THREE MEN, SIX LIVES*

"*Quantum Spirituality* is the most complete merger of science and ancient wisdom I have ever encountered. The result is the stuff of dreams, a magical story of the origins of humanity and our place in Creation. It reads like a visitor's guide to the Matrix, covering all angles—the spiritual, the scientific, and the psychological."

BETSY CHASSE, CO-CREATOR OF THE FILM *WHAT THE BLEEP DO WE KNOW!?*

"Peter Canova's book *Quantum Spirituality* is not just another take on the supremacy of consciousness. Solidly anchored in Gnostic Christianity, Jungian psychology, and cutting-edge science, it is a worthy addition to the rapidly growing body of literature that is reintroducing contemporary culture to the Perennial Philosophy, couching it in modern language and thought patterns. This is an important book that, although thoroughly accessible, needs to be read slowly and carefully digested. It deals with nothing less than the very explanation and meaning of life in our perception realm."

JIM WILLIS, AUTHOR OF *THE QUANTUM AKASHIC FIELD: A GUIDE TO OUT-OF-BODY EXPERIENCES FOR THE ASTRAL TRAVELER*

"*Quantum Spirituality* is a riveting, epic saga. It is a grand tale of our creation told in a mythic, spiritual, and scientific story that enlightens the reader about the mysteries of human existence. Its pages are rich with startling information that exposes the illusions of commonly accepted reality. The author merges spiritual insights derived from ancient mystics with modern quantum theory to create highly plausible concepts that ring true to every mindset, from the spiritually inclined to the logically grounded. *Quantum Spirituality* will provide every reader with life-altering knowledge essential to propel their own soul journey."

SCOTT CARLIN, FORMER PRESIDENT OF HBO DOMESTIC TELEVISION DISTRIBUTION

QUANTUM SPIRITUALITY

SCIENCE, GNOSTIC MYSTICISM, AND CONNECTING WITH SOURCE CONSCIOUSNESS

PETER CANOVA

Bear & Company
Rochester, Vermont

Bear & Company
One Park Street
Rochester, Vermont 05767
www.BearandCompanyBooks.com

Text stock is SFI certified

Bear & Company is a division of Inner Traditions International

Cataloging-in-Publication Data for this title is available from the Library of Congress

ISBN 978-1-59143-463-4 (print)
ISBN 978-1-59143-464-1 (ebook)

Printed and bound in the United States by Lake Book Manufacturing, LLC
The text stock is SFI certified. The Sustainable Forestry Initiative® program promotes sustainable forest management.

10 9 8 7 6 5 4 3 2

Text design and layout by Virginia Scott Bowman
This book was typeset in Garamond Premier Pro with Gotham Condensed used as the display typeface

To send correspondence to the author of this book, mail a first-class letter to the author c/o Inner Traditions • Bear & Company, One Park Street, Rochester, VT 05767, and we will forward the communication, or contact the author directly at **PeterCanova.com.**

Contents

Foreword

By James Redfield

This ambitious and fascinating book is a succinct message to us all.
Peter Canova seeks to resolve some of the most rivalrous debates in
history: those over the true meaning of human life and spirituality.

Even more impressive, he lays out a resolution that might just succeed. He has us look at two key spiritual philosophies that have been
staunch opponents for millennia. Through this ancient wisdom, we
experience a resolution, a new worldview reinforced by quantum physics and the depth psychology of Carl Jung.

But don't worry. This is no armchair intellectual speculation. Intellectual abstractions are not the endgame here. Peter wants us to
consider previously banned writings, important scientific concepts, and
necessary criticisms of religion to gain a deep personal experience of
higher consciousness based on our own dormant spiritual potential.
Right at the beginning, he gives us examples of his own transcendent
experiences as reference points to open us up to our own possibilities.

Peter fearlessly leads us into the battleground, first exploring the mystical world of Gnosticism, the historical "great heresy"
of Christianity, then comparing the contrasting concepts and principles
of the two views that shaped Western spiritual history: Christianity
and scientific materialism.

In the recovered writings and mythic tales of the Gnostics, we see a
model of humans striving to understand and seek an underlying spiritual
reality. The human journey in the Gnostic creation story is a heroic, individual, self-evaluating walk through the dense bog of materialism, finally

emerging into a higher dimension of spiritual connection, a connection subjectively experienced as an objective reality. The emphasis is on personal experience and the individual journey into transcendent mind.

Christianity, despite its early bravery, was hijacked by the aims of the Roman Empire. Unlike Gnosticism, it became a promise of the hereafter, not of the here and now, so it has emphasized obedience to doctrine and a collective waiting for the Divine to act. For Christianity, faith and patience were the determining factors in spirituality. For the Gnostics, the search for and the accumulation of greater knowledge through self-awareness would bring forth the spiritual connection.

From the beginning, the disagreement was fierce. The Gnostic emphasis on personal experience of the Divine was a threat to the Church hierarchy, so much so that they hunted down the Gnostics and burned their writings.

For Peter Canova, the loss of the Gnostic approach was tragic. And it was Christianity's loss, because centuries would pass before Christianity could shed its repressive institutional weaknesses and move toward a search for true unity with the Divine. Yet here, perhaps, we can begin to see a resolution of the paths, because, as it turned out, Christianity became stuck in the materialistic bog the Gnostics fought against.

Peter describes this impediment by pointing to the history of science, especially physics. For much of the past five centuries, science has fed the materialistic outlook—which tells us that reality is merely what the physical senses can detect—so from its inception, it claimed the mantle to explore the truth of human life and perception. The first major worldview it established was based on the theories of Isaac Newton, whose proponents began to push the whole idea of spirituality into the background. This old science told us we live in a world blind of nature, whose operations ticktock to mechanistic laws.

For centuries, science guided humanity's stroll into a scientific materialism that had been almost totally accepted by the twentieth century. This materialism affected everyone, including many Christians, eroding many core beliefs of Christianity.

At that point, the Theory of Relativity was introduced, which began to shake up this materialistic worldview. Simultaneously, quantum physics burst forth to have a similar effect, awakening both scientists and great numbers of individuals to a new sense that the universe was mysterious and even miraculous in its structure.

For Peter, quantum physics shows that the transmaterialistic inner-world journey of the Gnostics had validity. To emphasize the point, he also introduces Carl Jung's growth-oriented depth psychology. Jung pushed psychology to accept the idea that seeking wholeness and direct experience of the transcendent is an important human need. He also discovered the phenomenon of synchronicity, the experience of "meaningful coincidences," and declared this force to be an operating principle in the universe, designed to bring us moments of clarity, and sometimes direct help, in life.

Jung's work inspired a whole generation of seekers to pursue the wholeness of spirituality, including the experiences of aloneness, forgiveness, redemption, intuitive knowing and guidance, and even agape love—love without an object but as a higher emotional state. Most importantly, Jung's innovations indicated what happens if we deny our need for wholeness.

We repress this need with material pleasures and preoccupations, some of which can become inhuman. In his analysis of the sociopath, Jung showed how some individuals pursue power and control as an antidote to their lostness. He predicted that they could gather together in massive organizations, held together by a kind of mad, viruslike groupthink. He even warned that Hitler could be just the beginning. Such organizations, in the form of multinational corporations and foundations, could seek tyrannical power in our era as well.

In light of this situation, Peter's theme is timely. We all contain a spark of the Divine or the Master Consciousness behind the creation. We seem to be wired to seek spiritual reconnection and to discover a synchronistic journey that will lead us there if we pursue our journey in earnest. Here we may find the resolution toward which Peter is pointing.

Christianity emphasizes hope and faith in things unknown or in a hereafter described to us by others. Gnostic wisdom, by contrast, emphasizes direct personal experience of the higher consciousness behind all things as an inner realization in present time. Peter indicates how we can succeed in this journey to find these principles, dimensions, and realities for ourselves and have them operate in our lives.

This book outlines a journey that is critical for our evolution and could accelerate our awakening—a journey that could make all the difference in our lives.

JAMES REDFIELD is the author of a number of critically acclaimed books, including the international bestseller *The Celestine Prophecy, The Tenth Insight, The Secret of Shambhala, The Celestine Vision,* and *God and the Evolving Universe. The Celestine Prophecy* spent more than three years on the New York Times bestseller list and was the world's No. 1 bestselling work of fiction for two consecutive years.

Consciousness, Hidden Dimensions, and You

Something has come back from the grave to which it was consigned over two thousand years ago. It was buried away by the hands of men bent on suppressing perhaps the greatest secret of all time. It was not a demon. Neither was it the ark of the covenant. It was a virtual prophecy so powerful that, in the right hands, it could provide the key to the Holy Grail sought by modern quantum science—solving the mystery of Creation itself.

This ancient wisdom is mysterious. Viewed in isolation it would be hard for most to understand, but it just happened to be the missing piece of a larger puzzle. Like a fantasy movie where several keys must be recovered to open the magic door and unleash the power of the higher forces, this ancient bit of puzzle fit snugly between other already discovered pieces. Now the picture was complete. Everything meshed together. The entire puzzle revealed a striking picture, and it unleashed information that could transform the world.

The missing piece was a body of ancient mystical wisdom. The surrounding pieces awaiting its arrival were sciences such as quantum physics, biology, genetics, and psychology. Together they describe how Creation operates in a way that no single one of them could do alone. For the first time, we have all the information necessary to accomplish virtual miracles in our lives, control our fears, and begin to approach the mastery of sages past, like Jesus of Nazareth.

This is Quantum Spirituality, the merging of modern science and ancient wisdom. You may ask, why haven't I heard about any of this before?

Well, let me tell you the story of its obscure beginnings. It started in Egypt in 1945, the year the first atomic bomb was dropped . . .

Egyptian peasant brothers approached the limestone caves in the desert, poking the ground with sticks. They were on the outskirts of the town of Nag Hammadi in the lower reaches of the Jabal al-Tarif, a mountain riddled with numerous caves. Some of the caves were natural, others were carved into the mountain or used as ancient graves dating back over 4,000 years.

"I still have not seen any sabakh,*" one brother said, referring to a type of soil useful for fertilizing crops.*

"Look here," another brother, Muhammad 'Ali al-Samman, called out, pointing to something near the mouth of one of the caves.

As the others approached, they saw an earthenware jar about a meter high. The men stood for a while, studying the urn.

"Let us open it," one of them said.

"No!" Muhammad replied. "We may release an evil jinn from captivity."

This caused them to pause. "You know what this place is. This could contain gold," said one brother.

The men argued for a time, but then Muhammad raised his mattock and struck the jar. It was neither gold nor a jinn, but thirteen leather-bound papyrus books written in Coptic that tumbled from the shattered vessel.

Examining the books, one brother said, "These are for Christians. We have nothing to do with these."

Disappointed but still hopeful, Muhammad 'Ali al-Samman took the books back home, thinking they might yet profit from them in some way. He dumped the books and loose-leaf pages of papyrus onto some straw on the floor. In the ensuing weeks, the brothers' mother, Umm-Ahmad, would burn much of the discovery as kindling for her oven.

But something else burned in the al-Samman household—the desire for revenge. A man named Ahmed Isma'il had killed the brothers' father. Their mother urged them to take revenge on the man who had widowed her, and this the brothers did with a grisly vengeance. They dismembered Isma'il, hacked out his heart, and reportedly devoured it in a gruesome act of blood revenge.

The police started investigating the murder. Fearing they would search the family's home and find the books, Muhammad asked a local Coptic priest, al-Qummus Basiliyus 'Abd al Masih, to keep the texts. The Copts were people closest to the original Egyptians, who adhered to Christianity after the Muslim Arabic conquest. A local history teacher, Raghib Andrawus, somehow caught wind of the books and, upon seeing them, suspected they might be of value. He obtained one from the priest and sent it to Cairo.

Some of the texts came into the possession of Phokios Tanos, a Cypriot antiquities dealer, and then under the eyes of Jean Doresse, a French Egyptologist. Parts of the Nag Hammadi discovery eventually made their way to America and Europe, where wide-eyed scholars began to recognize that these were the original writings of the fabled Gnostic masters.

The Gnostics were the original Christian mystics who claimed to possess the keys to understanding the secret teachings of Jesus. Much evidence supported the existence of such teachings, including explicit passages from the Bible (which we'll discuss in detail in chapter eight).

I use the word *fabled* because the Orthodox Catholic Church, fearing Gnostic teachings would undermine the authority of their priesthood, completely destroyed the Gnostics and their works. They were so thorough that, before the discoveries at Nag Hammadi, the primary understanding we had of the Gnostics was through the writings of their orthodox persecutors.

The fact of the gospels having been hidden away in graveyard caves is evidence of the systemic suppression of the explosive knowledge they contained. The prevailing theory is that monks from the nearby Orthodox monastery of St. Pachomius hid the texts after orders from Athanasius, the bishop of Alexandria, gave stern warnings against any Gnostic influences.

Yet, now, thanks to the (random?) discovery and bloody actions of some Egyptian peasants, a 2,000-year-old genie had indeed appeared out of the jar—but not of the kind the brothers had feared. An ancient wisdom teaching branded as heresy by the Orthodox Church had been loosed upon the modern world—not to its detriment, but to its great benefit.

Astonishing information jumps from the pages of the lost gospels, information that eerily reflects modern theories of quantum physics and insights of modern psychology concerning the Creation, the hidden nature of reality, and the origins of consciousness. This information supports what has been called the "Perennial Philosophy," the common threads of mystical wisdom that run through every culture in every age, which can be summarized as follows:

- A Master Consciousness exists that is the source of every object and dimension visible or invisible, including our material world. In pop-cultural terms, we call it "the force."
- Our personal consciousness is a part of this Master Creative force.
- This force seems concealed from us, and there is a reason.
- The solid, 3D world has no objective reality independent of our consciousness observing and creating it.
- We are living in a simulated matrix, like a holographic illusion.
- Anyone can use the knowledge of these things to benefit their life.

HOW CAN THIS BOOK HELP YOU?

The statements above are pretty incredible, yes? Even counterintuitive to our everyday experience. However, if you think these matters are too fantastic or abstract to be useful, this book will make these issues less abstract and highly useful. The force has secrets to be revealed, and solving the toughest mysteries yields the greatest benefits. The most surprising thing you'll find is that all the wisdom we need lies sleeping inside of us. As Jesus said, ". . . nor will they say, 'See here!' or 'See there!' For indeed, the kingdom of God is within you."[1]

Each of us carries the story of the universe inside us. Our task, therefore, is to remember and consolidate the knowledge we already possess in the recesses of our being. So, here's what we're going to do. We'll take ancient Gnostic mystical wisdom and combine it with understandable quantum physics and psychology. You will see how they've all been saying the same thing about hidden forces and dimensions, despite 2,000-year gaps. This will give you a road map to gain some valuable leverage on life.

After adding the sum of these parts, I'm going to demonstrate that a Master Consciousness (you can call it any name you want) is the foundation of all things, including you and me. You'll see that we are important parts of this all-powerful, intelligent force but, like the central character in the science fiction movie *Dune,* we must first awaken the "sleeper within." Awakening this force will have world-altering effects on both you and the world at large.

As I write this, the coronavirus is devastating lives. Though it certainly has serious physical manifestations for some, fear and hysteria will end up harming far more people, lingering like nuclear fallout, long after the virus is under control.

I have no cure for viruses, but once we remove the clouds of mystery and complexity surrounding our existence, we get an astonishing view of hidden realities. This will help us navigate our lives with less fear and better decision-making in a challenging world. Such insights give us a sense that, when we face life's greatest trials, we're going to come out just fine—and that's not just a statement of faith. Once you're tuned in, you step beyond faith to *know* this statement to be true. This book shares the reasons why this is true from a scientific as well as a metaphysical standpoint.

This book will lift curtains to understanding our existence and help you gain more control over your life. It's about feeling more on top of life rather than being on the bottom. It will help you tap into sources of greater creativity. It will enable you to play life smarter, which leads to happier, more fearless, and more fulfilled lives.

Most people think reality is what the five senses can detect, but that's a very limited view. We're going to look beyond the apparent because if you slow-trot through life thinking it's one way but it's really another, you're working against the flow, not with it. To take advantage of life, we need to acquire knowledge about forces beyond the five senses. Therefore, we're going to look beyond superficial reality.

HOW THIS BOOK IS STRUCTURED

Unlike most spiritual or self-help books, the sacred mystical wisdom you'll learn here is grounded in scientific and psychological theory. Apart from introductory chapters addressing issues like the pitfalls of materialism, this book deals with five basic concepts:

1. Background on the Gnostics and the sacred feminine, particularly concentrating on the Gnostic Creation story.
2. Gnostic teachings in light of the modern psychology pioneered by Carl Jung.
3. The basics of quantum physics and its prominent theories.
4. Startling parallels between quantum physics and Gnostic mysticism, which give a complete description of our hidden reality as opposed to the virtual reality in which we live.
5. Finally, I'll tie everything together to make this information personally useful for you, showing practical ways you can align with the force of higher consciousness and tap into its life-changing information.

You're about to begin a journey to unravel the puzzle of life step-by-step, in an adventure that would make Indiana Jones envious. In the end, you'll discover something more valuable than legendary artifacts like the Holy Grail or the ark of the covenant. Instead of placing your faith in inanimate relics, you'll find a vast living power within yourself.

This is the promise and story of *Quantum Spirituality,* so enjoy the ride and let's begin.

1

Getting Started

Energy Matters

Einstein's famous equation E=mc² revealed that the matter on which we are so fixated comes from and is interchangeable with immaterial energy—they are the same thing in different forms. Solid matter, including our bodies, is really a form of unsolid energy, so reality works like this:

energy > subatomic particles > atoms > molecules > objects

Furthermore, subatomic particles, the smallest units of matter, contain just a tiny percentage of mass or matter.

Approximately 99 percent of atoms are energy, light, and space. *Only 1 percent is actual matter.* Our seemingly solid world is made up of particles in motion, with mostly space in between. That's an accepted fact, yet we build our reality around the material 1 percent and largely ignore the 99 percent—the far vaster matrix of nonmaterial forces from which matter is derived. On the smallest scale, material particles don't even exist, only energy does. This alone indicates we're experiencing, and living in, a form of illusion—yet something must be projecting this illusion.

Imagine yourself born into a solitary room with everything you needed to survive, so that all you know is that room. But one day you learn that this is just one room in a structure with multiple other rooms. Wouldn't you want to see what's in those other rooms and how they might affect you? In other words, wouldn't you like to experience those other realities?

The notion that hidden layers of reality exist in higher-frequency forms has great supporting scientific evidence, but our five senses can't detect them. This notion is counterintuitive to what we see and experience. The information in this book will help us lift these veils and show why we need to start a journey to tear away the barriers between us and the underlying dimensions.

THE WORLD AS ILLUSION

What we've just discussed is merely the tip of the iceberg. Most people find the popular *The Matrix* and *Blade Runner* movies intriguing. Guess what? The idea that we live in a simulated matrix or virtual reality is not just the imagination of screenwriters or producers. People from mainstream scientists to Tesla's Elon Musk are taking that idea very seriously, and not based on abstract whims.

The Matrix and *Blade Runner* films were based on the ancient Gnostic mystical wisdom, as described in this book and supported by new findings in physics. Evidence indicates that our universe is formed from unseen dimensions. Our reality may be a cosmic holographic projection of higher dimensions, like the settings and characters on the holodeck of the popular *Star Trek: The Next Generation* TV series. If this sounds far-fetched, read on.

This book is about grasping the nature of reality, the origin of life, the purpose of creation, and how you can navigate your path through it. If that still sounds too abstract for you, consider the following: the world's most effective thinkers have a deeper philosophy or view of life beyond just basically existing. Once you begin to comprehend how reality might truly work, once you better understand the source of your consciousness and grasp the big picture that is Creation and your place in it, think how much less fear and anxiety you will experience, and more control you could exert over your life. You can reach a point of knowledge and practice where you develop an inner confidence, like having an invisible army supporting you. The pieces of any puzzle you're facing then seem to fall into place in ways you could never have imagined before.

What's important here, and what this book attempts to do, is to help you establish and grow your own knowledge base. This framework of wisdom will enable you to work with the forces that can propel you to the highest levels of achievement you can attain in this life and beyond.

THE SPIRITUAL JOURNEY

This personal growth process is like building a never-ending skyscraper, where you forge a foundation of knowledge then build stories upon it by accumulating new levels of awareness. Sometimes, as in many complicated construction projects, you have to go back and make revisions. What you may think of as the truth at any given time can be replaced by newer, higher revelations or information. I'm speaking of an open-ended learning experience.

Here's how religion and spirituality differ: Religion usually claims to possess *the* truth, as if truth could be contained in a box. Spirituality, by contrast, is an open-ended exploration, a constantly unfolding subjective journey where each of us arrives at some view of the truth through different doors. Religion beckons the traveler to stay in the confines of the church, synagogue, or mosque, but the spiritually minded person takes what he or she can, says thank you, then moves on to the next waystation along the path, because truth is a journey, not a fixed stop.

CONSCIOUSNESS BEGINS WITH YOU

As a voyager of knowledge, you are going to learn about the nature of consciousness and the role it plays in the creation of reality, from both mystical and scientific perspectives. Here is where you can start to look at practical uses for your mind, to create a better life for yourself.

Know thyself.
DELPHIC MAXIM

As it is above, so it is below.
HERMES TRISMEGISTUS

Within these two ancient wisdom sayings lie the keys to understanding the nature of reality, both our personal reality and the wider creation we experience around us. If consciousness permeates everything, we can trace its operation within us and see that we are truly smaller reflections of larger forces that created the universe. We mirror the vast energies that shaped the great stage we call life, and consciousness is the stuff from which that stage is built.

WHERE DO YOU WANT TO LIVE?

You have two choices in life: You can live in the basement, going with the flow of your daily routine, having limited comprehension of the world around you. Or you can build that skyscraper previously described, and develop a higher, more refined perspective to view the full panorama of life.

Living in the world of the five senses is like seeing the tip of the iceberg and ignoring what lies beneath. That's exactly what sank the *Titanic*, and such ignorance can sink you, too, in so many different ways. The Gnostics said something over 2,000 years ago that Carl Jung and modern psychology have only recently recognized:

> If you bring forth what is within you, what you bring forth will save you. If you do not bring forth what is within you, what you do not bring forth will destroy you.[1]

The ancient mystics knew, and quantum physics is now discovering, that our world is but one link in a chain of dimensional existences. Ours is quite likely the lowest and most ignorant link in that chain, which, in the worst-case scenario, from several perspectives can be experienced as hell. But, if the secrets of Creation lie sleeping within us, we can break the chains of illusion by working with consciousness to open up new vistas.

THESIS

I refer to the central premise of this book as a thesis. A hypothesis is an assumption made for the sake of testing before any research has been completed. A theory, on the other hand, explains phenomena already supported by data— a statement or theory. A thesis puts forward propositions or statements developed, supported, and explained by means of examples and evidence. This book takes quantum theories with established experimental data, then cross-references and validates them with psychology and mystical wisdom to demonstrate the existence of higher realities. Here is the book's stated thesis:

- Objective reality does not exist except in the mind of an intelligent Master Consciousness that manifests Itself through multiple frequencies, like a radio signal broadcasting through different bandwidths.
- This Consciousness operates through the transmission of light energy, and emits an energy field called the quantum wave potential in physics.
- The energy field is the raw ingredient that Consciousness draws upon to form different interpenetrating realities or dimensions of existence.
- Like a stack of pancakes forming the totality of creation, these dimensions support and affect each other in descending levels of denser vibration, down to the matter of our perceived physical world.
- Personal human consciousness is part of the creative Master Consciousness, albeit operating in a greatly limited frequency of awareness that makes us think we're independent operators, separated beings from the Whole.
- Human beings are the bridge between the material world and the higher energetic dimensions of consciousness, but we must awaken to this reality to activate our innate co-creative power.

THE FRONTIER OF DISCOVERY

We'll discuss the properties and operation of Consciousness and the energy field in coming chapters, but you may wonder why more people aren't aware of this information. Pioneers once pushed the boundaries of the American experience outward, onto the vast Western frontier in a movement that would eventually shape life on the entire continent. But the information from the edge of the frontier took time to reach and be absorbed in the older, settled areas back east.

We're now talking about a frontier far bigger than the Old West—the greatest frontier of all: the quest to understand the Creation and our place in it. The primary instruments for conquering this frontier are not wagons, horses, or guns, but *information*. Assembling, absorbing, interpreting, and collating information into a pattern that reveals the mystery of life is the task of our modern pioneers, and such information takes time to disseminate and interpret.

THE FORCES IN OUR LIVES

This book is about the *forces* that shape human experience. Whether you are a believer or a materialist, a spiritualist or a psychologist, no one can deny there is a pattern of forces that affects our lives. This world is full of challenges. Our lives are defined by the degree to which we overcome our challenges. Life is a story of our desires and the opposition to those desires. Opposition may come from other people or from circumstances, but I have never met a person who did not face obstacles, adversity, and suffering in their life.

In the realm of dualities, our desires naturally give rise to things that oppose those desires. This happens so that we can grow through experience. At the end of the day, life is really a personal school, a boot camp we attend for growth. Our human journey is a voyage of discovery—it's the stuff of art, the core of religion, and the object of scientific inquiry. This book is meant to help you on your journey.

Within these pages lies a pathway, a yellow brick road with signposts along the way that will help you chart your own course to begin your journey and gain some control over your experience, so the trip gets less stressful. In a true irony of cosmic proportions, you will see how the salvation of science lies in our past, the salvation of spirit lies in our future, and the salvation of humanity may lie in the marriage of science and religion, reborn to common insights about our origin, destiny, and purpose.

So, my fellow travelers, Bon Voyage! and let's continue the journey that might change your lives!

2

My Story

I Get a Startling Awakening

Growing up, I never had remarkable insights about life. Sports, girls, and homework were the norm for me, as with many young people my age. I was, however, rather introspective. I lived a lot of the time in my head—maybe too much sometimes. I had an undercurrent of alienation, a feeling of weirdness about life, always asking myself, What's this world about?

So, I responded when I saw an ad in the newspaper back in the 1970s (did I just date myself a bit?) about a course for developing ESP—Extra Sensory Perception—developing psychic abilities.

EYE OPENER

We went through a week of training using hypnotic induction, a mental state related to meditation. The final day was devoted to manifesting medical intuitive abilities. We paired off, one person reading from an index-card bank of people with medical issues. The only info we were told was the person's name, age, and location—then it was up to us to relate what we saw. Here's how it went:

Partner: "I have here Mr. X, age 74, from Miami Beach. What can you tell me?"

Me: [Pause] "I see his heart tilted at a funny angle, pinching off the aorta."

Partner: "What are you saying?"

Me: "He has a blockage in his aorta."

Partner: "That's correct. What else do you see?"

Me: "There's a foreign object in the bottom part of the heart—it feels rather metallic. I think it's a pacemaker."

Partner: "That's correct. Can you tell me what year the pacemaker was implanted?"

Me: [I see the hands of a clock spinning wildly, suddenly stopping at 10:00.] "I think it was implanted ten years ago, in 1965."

Partner: "Wow! You hit on all three questions."

Me: "You know, I almost dismissed the images in the same instant they flashed as pictures across my mind. I guess our first impressions are the best ones, if we let ourselves believe them."

Once I'd accepted the reality of what I was doing (and I did it over and over again), and stilled all the dismissive disbeliefs my mind cranked out, the floodgates opened. I let my rational mind take a breather to allow my intuitive mind to take control. For a while, I became the "psychic doctor" of the working-class Boston neighborhood I was living in as a business school student. Clairvoyance, clairaudience, psychometry, psychokinesis, remote viewing, and accurate premonitions were some of the phenomena that became part of my daily routine.

I was like a man who'd been tuning a radio all his life but only getting static, until one day a clear channel came through. I realized this channel had been broadcasting 24/7 since time immemorial, only it was not going to attune to me on its own—*I needed to attune to it.*

LIGHT BULB MOMENT

As I was lying in bed in a semilucid state one early evening, a realization came in the form of a brilliant light. It was more a feeling than an idea—*everything in the world is a living, interconnected whole,* and by "living" I mean threaded together by an Intelligent Consciousness. How else could I know what was going on in other people's bodies with only names to go on, unless we were all connected at some unseen level? It was as though information lay in some invisible storage vault waiting to be tapped—information that had great value in the "real" world.

That single realization altered the course of my life. I'd once gone through a period with huge business and personal problems stacked against me. Without going into the gory details, I was overwhelmed and it was hard to see a way out. Contact with this higher source of information didn't help on just that occasion—it saved my life several other times. I did come to realize, however, that psychic abilities can degenerate into parlor tricks unless directed toward the higher purpose of liberating us from ignorance, so I set out to learn about the nature of this Source that was influencing me with its universal information.

MY JOURNEY BEGINS

I devoured every book I could find on spirituality and psychic phenomena. I'm a Capricorn, so for me it didn't just end with the acceptance of these newfound abilities—I needed to understand the nuts and bolts behind them. My research eventually led me to the field of quantum physics. In retrospect, I realize why.

We are creatures of mind and matter, energy and spirit. We are conscious energy in physical form. Our being is molded by both subjective, intuitive experience on the spiritual, consciousness side and by physical experience on the material side. Spiritual wisdom explains physical reality from its source in nonmaterial dimensions. Quantum physics approaches reality from the opposite end, that is, observation of the resulting material world. Spirituality and quantum physics together form the intersection between unseen conscious energy and visible matter.

All this experience and research led me to another career parallel to my business life. I started writing books and speaking nationally to help people by describing what I had learned. I wrote a multi-award-winning fictional book series called *The First Souls* that was based on quantum and spiritual principles.

Quantum Spirituality is my first nonfiction book. I did not want to write another positive-thinking book. This book doesn't contain clichés about how to materialize the person or job of your dreams, nor is it a training book for wannabe psychics. I want to give you a practical framework of spiritual and scientific tools to encourage and enhance your own personal life journey. The book will provide insights and indicate directions to help you launch and navigate that journey toward whatever dreams you want to achieve.

MY STORY IS YOUR STORY

I am not exceptional or gifted, despite some of my experiences. The ability to tap into higher dimensions is not restricted to me or a select few. My story is your story; we're all just following different paths to the same destination.

Helpful qualities I do have are a deep thirst for understanding the big questions of life and the desire to help others (and in doing so, help myself). Factors such as a desire to serve and empathy toward others are terrific gateways for growing and helping ourselves.

The combination of this heartfelt desire and a healthy intellectual curiosity to better understand life was the right brain/left brain engine that propelled me along this new path a little more quickly. But all of us can focus and choose such a path by using our inherent faculties of will, imagination, and directed feeling.

This book is a tour de force of spiritual and scientific wonders. We will journey back into history to trace the origins of a universal spiritual wisdom, and move forward to see cutting-edge science rediscovering the ancient knowledge that lay suppressed, twisted, and buried by the retardant forces of darkness or ignorance.

And out of all this a triumphant narrative emerges, an unbreakable thread that tells the story of your life, your origin, your destiny, and your purpose. No matter what level of consciousness or belief you occupy in this lifetime, this story will provide a framework to formulate your own path of spiritual growth.

Use the information gathered here, gleaned from long hours of research, meditation, and insight, and incorporate it into your own life, your own practice. You can graduate from vague or abstract ideas about how life works to a practical vision based on illuminating knowledge. The goal is to transcend speculation, faith, and the misdirection of materialistic science to experience actual results from connection with higher consciousness.

Like in the movie *Field of Dreams,* if you build it (a place in your being), they (experience and knowledge) will come. If you build a feeling in your heart and a focus in your mind, and you persist, it will come . . .

3

The Materialist Worldview

Spiritually Limiting Perspectives

If you are going to embark on a journey of spiritual seeking, it's best to understand the ingrained traditions, both religious and scientific, that you will be up against, because it gets to the heart of whether we are here due to a higher intelligence or just an accident. When science and religion fall into orthodoxy, truth is the casualty. The word *orthodoxy* comes from the Greek, meaning "right-thinking," as in right, true, or straight opinion. This should send up red flags to any free thinkers or truth seekers.

As with the modern ideas of political correctness, wokeness, and cancel culture, traditional religion and modern science have employed the practice of stifling opposing ideas with an unquestionable standard or dogma that are the views of one group, tradition, or hierarchy. The price people have paid for bucking those established views ranges from social ridicule and professional ostracism, to violence and loss of life.

The fact that the Western world has existed in a relative state of spiritual retardation for the past two centuries can in no way be laid solely at the doorstep of the church and religion. Like so many out-of-balance things in life, when science carved up religion, we went from one extreme to another. In breaking the shackles of religious control, science decimated religion. But from a spiritual point of view, it ended up throwing out the baby with the bathwater.

Scientific laboratories and halls of academia are filled with elite corps of religion-bashing secularists whose core view of life is based on materialist outlooks that color what they teach and study. Though science is less violent than the traditional church in enforcing its beliefs, it still has a chilling effect on the modern mind when it comes to phenomena that its instruments cannot probe—namely, spiritual realities.

In the scientific world, this has resulted in struggle around the concepts of consciousness and the origin of life. Secular materialists, like biologist Richard Dawkins (listen to some of his interviews), tend to be supremely smug and self-assured of their correctness. If you contradict their creed that a chance occurrence of particles somehow led to conscious life, you are labeled as an intellectually deficient religious Neanderthal.

Each to his own—but the problem is that these people are shaping the minds of our children with very few balancing points of view. They've pretty much run the opposition off the campus, using peer pressure and political correctness. Finding professors or scientists who oppose materialist world views is harder than finding a conservative in Hollywood.

The outlook and premise of materialism spill over into every aspect of life. The failed socioeconomic policies of communism and socialism came to life from the wellspring of materialism, as did anarchism, nihilism, and atheism. All these "-isms" destroyed more lives along the way than the church did on its best medieval persecution day. If you're okay with these philosophies, fine—but they do not have an enviable track record.

DARWINISM

Much, if not all, of the materialist mindset finds its roots in interpretations of the work of Charles Darwin. Darwin began as a scientist who had some notion of God, but he evolved (pun intended) into an agnostic. However, I don't believe Darwin ever dreamed of his theories becoming the basis for a hardened sociocultural materialistic outlook that would color every aspect of our modern psyche and replace the idea of a supreme being with . . . nothing.

The passion with which Darwinists and Neo-Darwinists seized and extrapolated upon Darwin's work was ironically religious in its zeal. So much so that the real underpinning of modern Neo-Darwinism is more ideological than scientific. The fact is that, as a science, his theories are full of holes, and the Neo-Darwinists who espouse it are playing a smoke-and-mirrors game to promote a materialistic philosophy.

THE CHURCH OF DARWIN

Ironically, the zealotry of this theory gone awry was directly birthed by religion gone awry. Darwin's findings were largely used by a succession of scientists, philosophers, and social engineers seething with resentment against the shackles of a church that had stifled the Western world for centuries. Later in this book we will discuss how one narrow, literalistic vision triumphed over a more spiritual version of early Christianity, then married the Roman Empire, birthing the authoritarian behemoth called the Christian Church. History plainly shows the repressive nature of this institution and its judgmental, punishing God.

Darwinism became a great weapon in the arsenal of scientific materialism, aimed to discredit and shed the chains of a religion imposed by the biblical idea of an anthropomorphic God. That means a distant God that was a reflection of human flaws instead of God as a higher permeating benevolent force that is an essential part of us. While we can't blame Darwin personally, Darwinian observations have been colored and used to promote a pervasive, materialistic worldview with many characteristics of an ideology.

Like so many other revolutions, the rise of materialistic science broke some shackles only to replace them with different ones. Today the materialists rule the roost. They have a social and intellectual lock on academia in a way that would make the medieval church proud.

DARWINISM: SCIENCE OR IDEOLOGY?

Many people are unaware of how Darwinism morphed from the natural observations of one man into a pervasive cultural ideology. Let's examine the mindset and psychological roots that enabled this to happen, by looking at telling quotations from leading evolutionary proponents.

Biologist Richard Lewontin

> We take the side of science in spite of the patent absurdity of some of its constructs . . . we have a prior commitment to materialism . . . to create an apparatus of investigation and a set of concepts that produce material explanations, no matter how counterintuitive. . . . Materialism is absolute, for we cannot allow a Divine Foot in the door.[1]

George Wald, Nobel Prize Winner

> Spontaneous generation, that life arose from nonliving matter was scientifically disproved 120 years ago by Louis Pasteur and others. That leaves us with the only possible conclusion that life arose as a supernatural creative act of God. I will not accept that philosophically because I do not want to believe in God. Therefore, I choose to believe in that which I know is scientifically impossible; spontaneous generation arising to evolution.[2]

Professor Richard Dawkins

> Today the theory of evolution is about as much open to doubt as the theory that the earth goes round the sun.[3]

Some revealing information is conveyed in these quotes. A clear, deep resentment of religion and God is embedded in the psyches of these men—an apparent backlash against the centuries of church domination in human affairs. But, as you will see, in their haste to latch onto the liberation train they saw in Darwin, they traded one limitation on life for another.

We also see some remarkable admissions above, that Darwinist materialistic evolution is not a science for these people—it is a bias, and an authoritarian one at that. It is a predetermined commitment to a philosophy or disposition regardless of facts or evidence. Sounds a bit unscientific, doesn't it? But despite this, Professor Richard Dawkins consistently proclaims his indignation that anyone could ever possibly question the holy book of Darwin.

Darwinists claim that all organisms have a hypothetical common ancestor that, over a vast time period evolved and branched out into different species. This supposition is completely contradicted by the global fossil record, which indicates that bodily forms for all of today's advanced phyla appeared suddenly, more or less side by side, during the explosion of life throughout the Cambrian period, about 500 million years ago. Here is an excerpt by Professor Jay Gould, a self-proclaimed orthodox Darwinist:

> The extreme rarity of transitional forms in the fossil record persists as the trade secret of paleontology. The evolutionary trees that adorn our textbooks have data only at the tips and nodes of their branches; the rest is inference, however reasonable, not the evidence of fossils . . . in any local area, a species does not arise gradually by the gradual transformation of its ancestors; it appears all at once and "fully formed."[4]

Another issue of contention is that one species can mutate or evolve into another. One scientific/scholarly website by Stephen L. Talbott, devoted to alternate theories of evolution called The Third Way, (coincidentally, a term I use in my own work) says this:

> Neo-Darwinism ignores important rapid evolutionary processes such as symbiogenesis, horizontal DNA transfer, action of mobile DNA and epigenetic modifications. Moreover, some Neo-Darwinists have elevated Natural Selection into a unique creative force that solves all the difficult evolutionary problems without a real empirical basis.[5]

Darwin has this to say about his lack of evidence that one species can become another or that all phyla descended from a common ancestor:

> To the question why we do not find records of these vast primordial periods, I can give no satisfactory answer. . . . The case at present must remain inexplicable, and may be truly urged as a valid argument about the views here entertained. . . .

But I believe in natural selection, *not because I can prove in any single case that it has changed one species into another* [emphasis mine], but because it groups and explains well (as it seems to me) a host of facts in classification.[6]

At least Darwin is more honest about his theory's potential shortcomings than his adherents. I wonder what Mr. Dawkins, a high priest of Darwinism who claims no rational person should doubt his views of evolution, would say about evolution's founder questioning himself?

The biggest source of contention to Darwin's admirable work lies in the title: *On the Origin of Species.* It's a complete misnomer. Nowhere in the entire book is the origin of any species or proof of a common origin given; Darwin only addresses the evolution of species *already* in existence. He doesn't *prove* a common ancestor, he *infers* it by analogy:

. . . But analogy may be a deceitful guide . . . Therefore I should infer from analogy that probably all organic beings which have ever lived on this earth have descended from some one primordial form, into which life was first breathed.[7]

Ah, but Neo-Darwinists want to promote their ideology at any cost. The same professor Gould quoted above was not above a little creative editing to cancel any mention of a higher power by Darwin. Darwin's sixth edition of *On the Origin of Species* (1872) said: ". . . *having originally been breathed by the Creator into a few forms or into one . . .*" Gould cites the same passage but edits out the word "Creator," apparently playing God himself.[8] You see, any mention of a higher intelligence is a no-no for these folks. In any case, the closest Darwin comes to describing the origins of life is that something "breathed it into existence."

From this biblical-sounding genesis, Neo-Darwinists have extrapolated the "undisputed fact" that life and consciousness had their origins from evolved matter. Really? Where did Darwin say that? That dogma foisted on the public is disingenuous at best, yet careers have been ruined, jobs lost, and minds brainwashed. Ben Stein did an excellent film a few years back called *Expelled: No Intelligence Allowed,* which documents the harassment and persecution of academics questioning sacred materialist dogmas.

A fair commentary on Darwin would be that his speculations on the evolution within species seem valid. If his proponents stopped there, we'd have no argument. The problem is that they have portrayed his observations as a factual theory of life's origins. Darwin neither explains the origin of species nor even conclusively demonstrates how one species evolved from another.

As stated earlier, Darwin's theories on the gradual evolution of species and on one species evolving into another also face problems. The Burgess Shale findings in Canada in 1909, under the director of the Smithsonian, Charles D. Wolcott, indicate that no gradual evolution ever took place but rather all phyla or body plans occurred suddenly and simultaneously in the Cambrian period. Wolcott hid this fossil record in the Smithsonian for eighty years, seemingly because it contradicted dominant Darwinian beliefs regarding gradual evolution.[9]

The actions of Wolcott and Gould, as related above, clearly show the mindset of ideological zealotry trumping honest inquiry and foisting prejudicial views onto the public as facts. But let's assume everything Darwin (not his followers) said about gradual evolution is correct. Nothing he contended contradicts the proposition that consciousness, not matter, is the source and foundation of existence. In Gnosticism, Creation is described as a gradual *spiritual evolutionary process* beginning in nonmaterial realms with a descent of consciousness through different dimensions down to the physical world. This opposes biblical Creation stories—*ex nihilo* (from nothing), with a wave of God's hand in six days—an outlook despised by Darwinists.

Gnostic Creation indicated that all matter contained consciousness through the agency of Sophia, the World Spirit. In quantum terms, this represents the universal energy field permeating all things. This field reacts to a directive intelligence, and the actions of consciousness on this field produce an electromagnetic holographic representation of 3D reality. We will explore this holographic view in future chapters, when I show how quantum science and rediscovered spiritual wisdom are eroding the materialist outlook. But the point is that nothing in this view excludes single cells imbued with primal consciousness evolving into more complex organic structures. Such evolution, however, would not be random, but consciously developed.

The mystic view was that, at some point, spirit consciousness accelerated the development of existing physical forms to become vehicles capable of receiving and holding higher self-awareness. The fossil record and the sudden appearance of Homo sapiens support this possibility. Either way, there is a symbiotic relationship between aspects of the Darwinian view of evolution and the unfolding progression of consciousness in the material dimension. Yet, if it were up to people like biologist Richard Dawkins, views like those presented here will never see the light of day before ridicule drives them out of the auditorium.

Darwin never explained how inorganic particles led to life and consciousness, but his acolytes have expanded his findings into conclusions he never made. These post-Darwinian dogmas have been undeservedly elevated to the level of accepted fact by many academics, despite their gaping flaws.

In presenting my views on how preexisting intelligence could have created life, I present them as thesis, not a done deal. So why is Neo-Darwinism treated as sacred fact? Why are theories of creative intelligence not taught side by side and debated, so students can make up their own minds? Unilateral exposure and blind acceptance of pseudo-Darwinist theory constitute dangerous academic brainwashing.

If you think that statement is overblown or dramatic, consider the following: Merriam-Webster's Collegiate Dictionary defines *brainwashing* as "persuasive by propaganda or salesmanship." Therefore, brainwashing is a result of repeatedly feeding someone an ideology reinforced by authority figures, such as professors. How a person thinks about the origins and purpose of life, if only subconsciously, has a deep impact on how he or she approaches life in general.

MEANINGLESSNESS

The human mind is impressionable, especially at young ages. Look at the rigid prevalence of political correctness and cancel culture on campuses today. Accepting a mantra that life is a random accident of particles that happened to evolve on their own, the unconscious mind adopts that as an ingrained attitude of meaninglessness about life.

Indeed, philosophically, Darwinism rapidly became conflated with the materialistic concept of "meaninglessness," as it holds that life arose by accident without guiding intelligence. That being the case, we can use this outlook to justify nearly any behavior—good or bad.

Aldous Huxley, whose grandfather Thomas Huxley was an agnostic known as "Darwin's bulldog," explained the meaningless concept and the motivations of people who seized on Darwin's theories as a revolutionary social tool.

> The philosophy of meaninglessness was essentially an instrument of liberation . . . liberation from a certain system of morality. We objected to the morality because it interfered with our sexual freedom; we objected to the political and economic system because it was unjust.[10]

This statement by Huxley is critical because it gives insight into the social objectives of materialism and the forces that have been driving the prevailing mindset in the Western world. These influences produced conflicts that are still raging before our very eyes. Understanding this is so important, I devote the Afterword to explaining what's happening beneath the conscious awareness of most people, and how we can find the "Third Way," a path between the extremes of traditional religion and scientific materialism.

With that said, let's continue our journey.

4

The Perennial Philosophy

A Sacred Ancient Wisdom

From the dawn of history, a mystic universal wisdom tradition existed. It addressed the great questions of life: Who are we? Where did we come from? Why are we here? It was mystical because it was not derived from math, instruments, or scientific observation. It drew from subjective inner experiences with a body of information that lies beyond the physical senses.

- It was mystical because it dealt with matters beyond the five senses.
- It was universal because variations of it were taught in nearly every civilization on Earth.
- It's called wisdom because the seers who brought this information to the world were in constant and direct contact with a higher Source. That's my definition of wisdom—knowledge gained by directly experiencing something as opposed to imagining it. We're talking here of direct contact with a living higher Source of information.
- It is a tradition because it was taught and handed down as a living, organic transmission of intelligent light energy, from teacher to pupil. And then it was lost, at least in the Western world.
- It was, sadly and ironically, a particular form of Christianity that was responsible for breaking this chain of wisdom that existed throughout millennia.

This ancient, universal, and mystical worldview came to be called the Perennial Philosophy. The word *perennial,* meaning constantly recurring, highlights the fact that this wisdom tradition has been universally present in all cultures at all times, though often suppressed.

THE PRINCIPLES

Though it had variations from culture to culture, the Perennial Philosophy had common elements:

- Consciousness is the only thing that exists. It is energy that projects Itself as substance through polar forces variously called male and female, yin and yang, or positive and negative. The phenomenal material world in which we live is not fundamental but is the projected reality of this Conscious Energy.
- Light energy is the vehicle by which Consciousness manifests Itself. Light contains intelligence.
- Humans are multidimensional creatures. We have a true self, consisting of a mind that eternally exists as part of a Master Consciousness, and a projected ego-self of limited awareness, which resides in physical forms.
- No separation truly exists between human souls and God, or the Supreme Consciousness. The belief in separation is an error, a delusion of the mind that can be corrected.
- Using intuition, imagination, knowledge, intention, and faith in a balanced manner allows one to break out of space, time, and history to acquire enlightened information from higher levels of consciousness beyond space-time. Humans possess the capacity to transcend the ego illusions of time, space, and history, to reawaken to the higher conscious unity of all things by gaining knowledge through direct experience of their own hidden, higher state of being.
- The primary purpose of each person is to reawaken to the higher consciousness within and recognize that material experience is but an illusion or dream of our own making.

ORIGINS

The diagram in figure 1 traces the major expressions of the Universal Tradition through the ages. It is not strictly sequential. For instance, the Great Mother epoch would have overlapped with Shamanism. Similarly, the Gnostics were a particular branch of the Mystery Schools, so you could say the Mystery Schools lasted until 400 CE.

The tree below shows the principal forms Perennial Philosophy took over the ages and which form might have been more evident during a particular period.

SHAMANISM
Pre 12,000 BCE

|

GREAT MOTHER
Pre 3,000 BCE

|

MYSTERY SCHOOLS
1,500–1,200 BCE

|

GNOSTICS
500BC–400CE

CHRISTIANITY

Figure 1. A chart of the Perennial Philosophy in the West

Shamanism

Simply stated, shamanism espouses the belief that two worlds exist side by side, the seen and the unseen. The shaman was one who walked between the worlds. The shaman would bring back information from the energetic world, the world that was the light energy blueprint for the material world we know. This information would be the early nucleus of the Perennial Philosophy. The shaman used it to heal, advise, and guide people. The shaman was a "guider of souls."

The Great Mother

Prior to 3,000 BCE, many of Earth's cultures were matrilineal and worshipped some form of the Great Mother or Great Goddess. In Egypt, Isis represented the sun and her husband Osiris represented the moon. The light of the moon (Osiris-male), being a reflection of the light of the sun (Isis-female), indicated the superiority of the female. The name *Isis* means "seat," as in seat of authority. Isis was often depicted in ancient Egyptian symbolism wearing a throne/seat upon her head. During this period, mysteries began to develop around the feminine faculties of intuitive imagination and a feeling for the rhythms of the earth and nature.

Formalization

Moving forward, it seems that perceptions of sacred wisdom became systematized in the early Hindu tradition and carried west into Persian Zoroastrianism. The Alexandrian Greek conquest of Asia later opened the door for a melding of Greek philosophy and Oriental spiritual wisdom.

As evidence of the merging influences, the Christian polemicist Hippolytus observed in the early third century CE that: *"Brahmins in Alexandria . . . affirm that God is light. But not such as one sees by . . ."*[1] The presence of Hindus spreading their knowledge in the Greek city of Alexandria, Egypt, illuminates another chapter in the history of the Perennial Philosophy—the Mystery Schools.

A body of formalized schools, or centers of wisdom teaching, developed from India throughout the Middle East into Europe, particularly around the Mediterranean. These centers taught variants of the Perennial Philosophy that permeated the Greco-Roman West. This led to a fascinating blend of Greek, Oriental, and Jewish mysticism most highly embodied in a multicultural group called the Gnostics. It's to the Gnostics we'll turn to learn how ancient wisdom gave answers to life's mysteries that are being repeated today in science labs and psychology papers.

Suppression

In the diagram in figure 1, you see a break between Gnosticism and the Christian era, even though Gnostics were among the original Christians. We'll discuss the story of how the Romanized church destroyed classical Gnosticism later.

5

The Sacred Feminine

The Feminine Principle in Mysticism

In the previous chapter, we touched upon the Great Goddess as part of the chain of the Perennial Philosophy. Much of this book concerns the ancient feminine spiritual principle. It will become clear in our unfolding story that the feminine qualities of wisdom, intuition, insight, imagination, and feeling are the gateway to our higher natures.

The ancients said it was the feminine qualities of wisdom seeking and creativity that caused spirit consciousness to fall from higher stations in search of new life experiences of mind and physicality. Wisdom, a traditionally feminine-associated state, is at the heart of the Gnostic Creation story. The feminine force of wisdom is compelled to seek knowledge and experience, even if on impulse with ensuing consequences. That paradox forms the Gnostic story of Sophia and the creation of life as we know it.

Another good reason for focusing on feminine spirituality is that it has been so suppressed as to be almost forgotten, yet it was the highest legacy humanity possessed in an era that was in some respects more advanced than the present age, at least in terms of deeper spiritual understandings.

For some, the feminine principle today is misunderstood and confused with modern political movements. The sacred feminine *is not* feminism. While there is some relation in a broad sense, the gender rights movement is a superficial echo of a much deeper story.

THE EARTH MOTHER

The sacred feminine principle is the recognition that God expresses Itself in a female aspect as well as a male one. Although God is one, the Divine Unity expresses Itself as two fundamental aspects or polarities in all dimensions of existence, including the visible universe. These polarities are one energy with different vibrations expressed with different characteristics. From time immemorial, these aspects have been described as male and female.

Ancient wisdom tells us that the chain of events that caused the Creation was led by the feminine aspect of God. Many ancient societies held that the Earth was alive with the seeds of life planted by the Divine Mother, or Sophia, who became Gaia, or the World Soul. Most early cultures and religions focused on Goddess worship that was equated with nature or the concept of Earth Mother. Not too hard to understand. Primitive humans would notice that life—humans, animals, crops, flora—were all born from something, and men don't give birth. When you think about it, early humans probably didn't understand the reproductive process. All they saw were women birthing, so life would have been associated with the feminine by primitive hunter-gatherers and agri-based cultures, and worshipped accordingly.

MATRIARCHAL SOCIETIES

A matriarchal society is not necessarily run by women, but it's guided by feminine orientations. Some other traits of matriarchal societies include:

- Less emphasis on violence and more on coexistence
- Harmony with land and nature
- Consensus rule or governance
- Lack of hierarchies
- Goddess worship
- Prestige from contribution, not conquest

Much evidence exists as to the feminine orientation of early humanity.

It is, in fact, in the Gravettian epoch of the Upper Paleolithic age from 25,000 BC to 20,000 BC that the first manifestations of anthropomorphic art appeared . . . it was nearly exclusively women who appeared.[1]

Archaeology indicates that the seminal Western culture of Egypt was a matriarchal society before 3,000 BCE. The goddess worshipped in Upper Egypt was Nekhebt. Lower Egypt called her Ua Zit. Egypt was not only matriarchal back then, but matrilineal. Women were the legal heirs to the throne. The pharaoh had to marry the eldest princess, who conferred the crown to him as her regent.

Figure 2. The Goddess in Egypt

Starting a few centuries before 3,000 BCE, Indo-European invaders imported a more warlike culture, with male deities, that altered the old practices of Egypt. This triggered an apparent war, as recorded in

Egyptian mythology, between the old female and new male deities, with males finally gaining ascendance.

Ancient Hebrew history mirrored the Egyptian experience—not so startling perhaps when you consider the years of servitude that Hebrew tribes spent in that land. The Bible mentions four matriarchs: Sarah, wife of Abraham; Rebecca, wife of Isaac; and Leah and Rachel, the wives of Jacob. Additional evidence shows early Hebrew reverence for a female deity.

In a remarkable biblical passage from Jeremiah 44:16–18, Jeremiah threatens the Jews who were still living in Egypt at the time. It seems the male god Jehovah was mad again that he wasn't getting his due. In response to Jeremiah's demands, the Hebrew women rebuke him, saying this:

> As for the word that thou hast spoken unto us in the name of the Lord, we will not hearken unto thee. But we will certainly . . . burn incense unto the queen of heaven, and pour out drink offerings unto her, as we have done, we, and our fathers, our kings, and our princes, in the cities of Judah, and in the streets of Jerusalem: for then had we plenty of victuals, and were well, and saw no evil. But since we left off to burn incense to the queen of heaven, and to pour out drink offerings unto her, we have wanted all things, and have been consumed by the sword and by the famine.

THE GREAT SHIFT

I could cite a litany of scholarly opinions on how the shift from matriarchy to patriarchy occurred, but instead I'll offer you my own perspective. Seeds inherent in the Creation itself caused the world eventually to tilt against the feminine. When consciousness first fell into material form, the feminine force dominated the minds of our early ancestors, like babies first attached to their life-giving mothers. The world would have seemed magical to the early humans, as they intuited their way through life in pulsations with the rhythms of nature.

But now consciousness found itself embodied in physical form subject to all the laws and perils of physical existence. Wild beasts preyed on our ancestors, who needed to gather food and seek shelter. They had to begin to organize and analyze, to develop ways of understanding and controlling their environment, and those are left-brain, male qualities.

As humans became more remote and forgetful of the source of their conscious existence, their intuitive-mind powers diminished and they came to rely on more physical and mechanistic methods of looking at the world and controlling it. This led to more aggressive left-brained male-oriented strategies for living. Now their egos got firmly into the driver's seat, and reliance on feminine intuition to contact the Source of life diminished.

This way of thinking as a survival mechanism eventually came to dominate the human thought process so successfully that today we are practically destroying ourselves for the sake of dominance and power over the environment and other humans.

CHERCHEZ LA FEMME

This French expression means "seek the woman." This is great advice, with more meaning than the coiners of the term may have understood. The history of the Sophia story and the sacred feminine has a larger overarching symbolism. The feminine principle is about the intuitive, feeling factor inherent in all of us. I mention left brain/right brain orientations throughout this book, or the rational, analytical as opposed to intuitive, instinctual views of the world. In the modern age, we've largely flipped the historical deck so the intuitive mindset has become subservient, even mistrusted, by the majority of people.

The sacred feminine in all its manifestations is a reminder that intuition, not analysis, is the gateway to higher consciousness. Though both approaches play important roles in life, contact with higher consciousness and other dimensions is essentially an intuitively experienced subjective process, possibly having an objective material result. This fact is reflected in the Gnostic understanding of Mary Magdalene. She was shown as the primary disciple who most intuitively grasped the radical mysticism of Jesus's teachings, while the other disciples struggled with their male rationality boxes.

Indeed, the Gnostic masters were so in tune with the feminine vibration that it opened the doors for them to bring down to this dimension the amazing information you will soon learn. So, in your personal practice you would be well advised to "seek the woman." Learn to develop—and just as importantly, listen to—your inner intuition and guidance. As the song says, "That's what it's all about."

6

Gnosticism

A Mystical Tradition Recovered

The spiritual content of this book focuses on Gnostic wisdom for two reasons. First, Gnosticism was the culmination of major Eastern and Western spiritual philosophies formulated over the ages by sages and masters. Secondly, this synthesis produced the most astonishing information found in ancient annals about consciousness, reality, parallel dimensions, and the Creation mystery as a whole.

The Gnostic's insights were so penetrating, you could say they were the quantum physicists and master psychologists of their day. We'll explore how their wisdom anticipated many theories of modern psychology and quantum science. The term "Gnostic" comes from the Greek *gnosis,* which means "knowledge"—but not knowledge in a mere rote, factual sense. It means knowledge derived from direct experience, in this case from contact with a higher source of consciousness.

The Western Gnostics thrived in the triangulated area from Athens to the Judeo-Syrian Middle East, then down to the great cultural melting pot of Alexandria, Egypt, where Greek, Jewish, and Eastern mysticism merged to address the great mysteries of life with razor-sharp focus.

The Gnostics predated Christianity. They incorporated, among other traditions, early Jewish Kabbalistic mysticism. But, as you'll see, the Gnostics eventually saw the secret teachings of Jesus as a supremely Gnostic revelation, and they became the earliest bloc of Christian converts.

LOST GOSPELS

We know about the Gnostics primarily through the recovery of their suppressed gospels from the desert sands of Egypt in the twentieth century, as I described in the introduction. They were all but eradicated by the Orthodox Christian Church, a story I'll chronicle in subsequent chapters.

WHY RECOUNT THE GNOSTIC
CREATION STORY?

After reading the information in this book, some of you may ask, What good is it to study the Gnostic Creation story or spiritual mysticism of any sort? The short answer is that it describes how our reality operates and how you can align with the flow of spiritual/psychic/physical energies that govern our lives.

The most important kernel of spiritual wisdom ever written comes from the ancient Egyptian Hermetic tradition, and it says, "As it is above, so it is below." This means that the operation of unseen forces that govern higher planes of existence are reflected within our world and even within our very being.

How did humans ever invent a complex computer? The idea didn't come out of a vacuum. The functions of the hard drive, the memory, the storage, and the software are all reflections of how our minds operate processing, storing, recalling, analyzing, and replaying information. All information in turn ultimately comes from the master program that organized the entire Creation, including your consciousness and mine.

The reason Jesus and other spiritual masters told us to seek answers from within, rather than without, is the recognition that we contain the blueprint of all Creation within us. We are microcosms of the macrocosm. The knowledge we uncover to progress intellectually and technologically is not so much learned as remembered. Higher knowledge lies dormant within us, waiting to be uncovered, and those things you think you've discovered are things you've already mastered in another dimension of existence.

What this points to is a multilayered reality. Our physical existence is really the end product of higher-intelligent realities. Our minds and souls occupy places in each of these realities according to the limitations we have accepted in order to experience individualized forms of existence, such as separate physical bodies.

The short answer to why we should learn Creation stories is that "the Force" is real and understanding its operations, as the spiritual masters did, is the beginning of mastering life, ending fear, and achieving the potential of your dreams. It's not mumbo jumbo. It's both a spiritual mystery and quantum science, and that's what this book is about.

GNOSTIC TEACHINGS

Mystical Gnostic wisdom comes to us primarily in the form of myths. By myths I do not mean fantasies but Joseph Campbell's more insightful view of myths as stories embodying universal truths in forms absorbable by the common masses. When speaking of subjects as deep as the origin of life and the universe, delivering that information as raw data would have been unintelligible to the common people of the ancient world. It's only in light of modern psychology and quantum physics that we can see how astoundingly accurate the Gnostic seers were in penetrating the mysteries of life.

The Gnostic gospels are full of rich imagery symbolically describing the forces behind the creation of both the material world and the energetic dimensions that preceded it. In the next chapter, I'll describe the central Gnostic myth of Sophia, but for now, here is a general outline of Gnostic beliefs.

GNOSTIC ORIGINS OF CONSCIOUSNESS

Gnostic texts describe the beginning as a still, unknowable depth (Gk. *Bythos*), which can also be interpreted as the foundation—unknowable, unfathomable conscious potential in the process of becoming self-aware. It was called Forefather by the ancients (Gk. *Propater*), or simply, God. One can use many other names to describe this consciousness—the Monad, the One, the Source, the Supreme Being, the First Principle, etc.

It is a unity containing two aspects or characteristics—a passive, potential side, which is substance, and an active side, which is thought. The passive aspect was called male and the active aspect female. You could say it was both Everything and Nothing, because nothing else but Itself existed. It was a static existence until It moved within Itself.

Its active side produced a thought that stimulated Its passive—substance. The interaction of these polar forces produced a movement of kinetic energy where the formless Nothing produced Something out of Itself, but Something of a different energetic frequency.

Now there was a Nothing and a Something. The Something(s) were called spirits, or mind-sons and mind-daughters in spiritual lore. They were new individualized projections of the One Consciousness, new points of intelligence. The appearance of new points of consciousness represented the transition from a passive to an active, manifested existence. This transition was the Primordial Unconscious (God) becoming self-aware.

SELF-AWARENESS

This might seem a bit dense, but think about it for a minute. If the Source is everything that exists, It has no points of contrast by which to compare or define Itself. It can't be fully self-aware. If, as an infant, you were plopped down on a barren island in perpetual darkness and somehow managed to grow and survive, how much self-awareness would you develop with no other humans by which to contrast yourself? Our awareness of ourselves is largely through the presence of other beings and objects, when we really think about it.

So, the Source projected other points of consciousness besides Itself, but here was the trick—these new points of consciousness were limited. They had to be limited in order to have a sense of individual identity. Why is this so?

I have an electric bike. Though the bike is capable of going much faster, the law doesn't want these vehicles to go over 20 mph, so they come with a built-in governor that limits the bike's speed, so I can never experience the bike's ultimate capability. If the new points of consciousness weren't limited, and they could achieve the ultimate infinite consciousness of the Source, they would just be reabsorbed by the Source, and poof—there goes their individuality. They need to have a different frequency of consciousness to perceive themselves as separate beings.

In traditional Christian terms, the process described above was the passive *transcendental* God-beyond-existence manifesting Itself from the unknowable depth to become the active *immanent* or knowable divine presence capable of being felt or imagined in varying degrees by others. Think of it as God now *within something,* as opposed to God *outside of everything.*

DUALITY/POLARITY

Now, instead of the single presence of the Source, something new appeared on the cosmic landscape. Other points of consciousness, or spirit beings, manifested from the mind of the Source. The Gnostics called these beings Aeons, Greek for "Eternities." The Hindus called them Mind-Sons and Mind-Daughters. The Aeons possessed differing levels of consciousness, or energetic frequencies. This variation was necessary for creating individual identities distinct from one another, similar to the way we all have different fingerprints despite all the human traits we share in common.

So, in the higher dimensions, individuality is determined by different levels of consciousness. All spirit entities, whatever their level of consciousness, were aware of their connection to each other and to the Source.

The Creation was a movement from the unitary existence of the Source to a state of duality. Instead of the One, there appeared more than one. Now a community of spirits existed that were projected by the Source. The image on the left in figure 3 on page 42 is the Nothing/Everything (God), a singularity. The image to the right shows how two states now exist, the Source *and* the individual community of spirit consciousness. This dual existence is where things *appear* separated, not

Figure 3. Unity versus multiplicity

Singular Existence
THE SOURCE

Dual Existence
SPIRIT CONSCIOUSNESS

unified. Note that I said "appear separated." The perception of duality is an illusion, since the spirit individualities are really manifestations of the Source, in states of limited consciousness. This limited awareness allows them to perceive themselves as separate entities when they are actually part of the Source, but in a lowered state of consciousness.

The newly projected beings contained the same internal mechanics and polar composition as the Source. Each was a yin-yang, male-female composite in the image pattern of the One. What is the purpose of the polarities? Think of the positive/negative polarities in the batteries of your flashlight. The interaction of those polar charges is what enables the flashlight to shine.

The interaction between polarities—attraction and repulsion, positive and negative—is what produces the light, motion, heat, and friction to make energy, which is the engine of creation.

In science, the action of negatively charged electrons moving around the positively charged atoms produce the light by which we detect physical objects and experience the material world as a reality.

We see polarities operating throughout nature and science—the yin and the yang, the positive and negative, the male and the female, the 0 and the 1 that make our computerized electronic communications possible—which are all manifestations of this polar duality.

NOT CREATION, BUT EMANATION

It's important to understand that God, the Supreme Consciousness, did not *create* anything in this movement toward manifestation. A creation is a thing fashioned by a maker but not of its maker's substance. The toy-maker Geppetto made the puppet Pinocchio like the artist making his sculpture in figure 4a, but it was not of the same substance as Geppetto.

God, on the other hand, *emanated* or projected portions of Its own conscious substance outward. Like you might see a bubble forming off another bubble, all conscious beings emanated or *flowed outward* from the Source and were of the *same substance* as the Source, as depicted in figure 4b.

Figure 4a.
To Create Is to Make

Figure 4b. To Emanate Is to Project

I really want you to grasp the Gnostic concept of how the Nothing or Source projected the Something to create an apparent duality, so here's a bit more explanation and illustration. The projections of conscious spirit beings from the Source were achieved through *conscious energy projecting Itself,* the way you might project a shadow copy of your hand on a wall by holding a flashlight behind it. The light energy propels an image of the source—your hand in this case.

This is a helpful analogy because you know the shadow of your hand on the wall is a paler reflection of your actual hand, just as the spirits were limited or diluted points of consciousness from the Source. In quantum science, particles and energies are distinguished from one another by their energy signatures, or frequencies. Pop culture calls these "vibrations."

So, the appearance of beings both visible and invisible is ultimately a matter of consciousness or intelligent energy projecting Itself in varying vibrations or frequencies. The variation in energy frequencies is what distinguishes conscious beings from one another, and creates the appearance of separate multiplicities in all material phenomena.

COSMIC ILLUSION

Despite the appearance of separate identities or points of consciousness, everything is still part of the Whole. It isn't perceived that way because the Supreme Consciousness "dumbed Itself down," so the limited parts could perceive themselves as individual beings, creating a sense of individuality through a kind of cosmic amnesia.

If this seems strange, think about it. How can an entity that is *everything* create *something* other than Itself? Well, it really can't except by projecting Its consciousness in limited doses. The limitation of consciousness creates a sense of individual identity. If the parts possessed the consciousness of the Whole, they couldn't be parts—they would reabsorb back into the Whole.

If the material world, for instance, did not exist in a different and limited state of consciousness or energetic frequency, it would be absorbed back into its ground energy state. You can see now why traditions such as Gnosticism and Hinduism say the Creation is a form of *Maya*, or illusion, the illusion being the belief that we live an existence apart from the Whole that projected us.

So, individuality is really reduced energetic frequencies of limited consciousness and awareness, creating a sense of separation from the Whole. This is what our lives are about—everything we experience, the good and the bad, the sublime and the profane, our triumphs and tragedies, all stem from this flowing down of the Eternal into lower illusory dimensions of mind and matter.

And this leads to a lot of trippy but exciting stuff as we peel back the veils shrouding the mysteries of the Creation.

7

The Great Gnostic Myth of Sophia

An Incredible Take on Creation

The Gnostic view of Creation began with the primary male-female Source or God. God emanated the first being, another polar entity called an Aeon. Think of Aeons as vast currents of intelligent energy occupying dimensions defined by vibrational levels of consciousness.

Remember how we discussed the rise of individual beings as limitations of God's consciousness? The Source used a reproduction pattern for the Aeons similar to the formation of human generations. The first Aeon generated another Aeon, and that one another, and so on. The Aeons interacted between their own internal male-female energies to reproduce. But the Aeons are currents of intelligent energy, so each successive Aeon was a whole dimension unto itself, differentiated by frequencies and aware of one another but also distinct from one another.

Each Aeonic level generated a new lower-level Aeon several iterations on down the line, like making successive copies of CDs. The original CD has the highest clarity, or fidelity in musical terms. Each copy of a copy degrades from the original clarity to some degree. So, the later Aeons were somewhat less conscious than their predecessors, though all of them were powerful divine spiritual beings.

Another way to think of Aeons is as mind-generated thought-forms, or ideals from the mind of God. Just as we display our own unique personal ideals or qualities, the Aeons exemplified various goals, visions, and desires of God. These divine ideals are called archetypes. An archetype is a blueprint of a basic pattern, way of thinking, or being. The Aeons were intelligent energy centers personifying essential ideals like Truth, Grace, Silence, Beauty, Mind, Thought, Faith, and Love.

Notice that these things are abstract ideals on some levels, but they translate as deeply experienced feelings or impulses that give basic meaning and purpose to our human lives. The Gnostics believed that our ability to conceptualize and be guided by these ideals radiates from God through the felt influence of the Aeons operating in the higher dimensions.

So, the Aeons, in modern social media terms, are major influencers whose effects pour on down to our unconscious psyches through the parallel dimensions.

Like the concentric rings of an onion, one Aeon generated another in a descending or outward projecting pattern around the Source, not as spaces or places, but as patterns of mind and energy. The spiritual panorama of this first Creation was bands, or dimensions, of intelligent energy expressed like radio frequencies radiating from a master signal differing in strength or amplitude.

A warning at this point—the story that follows will seem rather bizarre. Some of you may ask why I'm wasting your time with such a fantasy. But this "fantasy" symbolically embodies the wisdom of the ages, which laid the foundation for every major religion. I'll explain the symbolism and what the story is saying beneath the surface, to hopefully reveal some of the mystery behind Creation.

I also promise you—and this is the entire purpose of this book— that this story will come into startling focus once you see it through the lenses of modern psychology and quantum physics. If the Gnostics had had the technology and language of modern psychology and physics, the entire world would marvel at how people could gain such insights three thousand years ago. Instead, we're going to assemble the pieces of the puzzle using the tools of our age to solve this mystery. Anything of value worth having is worth working for, after all.

THE WAGES OF SIN (CHOICE)

The multiple levels of Aeonic conscious beings that flowed forth from the Source collectively formed the *Pleroma,* or Fullness—the Christian equivalent of heaven. The last, and youngest, of these Aeonic beings was Sophia, which is the archetype of wisdom.

The Aeons were aware of themselves as part of God but also aware of themselves as individuals. They felt the presence of the God that generated them, but the ultimate essence of the One was shrouded in an unknowable mystery, since if they fully knew God they would be absorbed back into Its Unity and lose their individual identity.

Of all the Aeons, the youngest or last-generated among them, named Sophia, burned brightest with the desire to know what could never be known, to fully know the unfathomable God. She was the feminine aspect of the Aeon associated with the quality of wisdom. Wisdom is driven to learn, so she attempted to rise and learn the ultimate mystery of the One, but she was restrained by the limiting consciousness boundary separating the unknowable Source from the Aeons. The way was barred to her.

The waves of passion and grief flowing from this realization attracted the attention of Authades (Gk. "Audacity"). Audacity is the principle of *opposition* that permeates Creation. It's important to understand the meaning and symbolism here. Opposition is that force which pulls beings away from the Unity to achieve individual desire.

The force of opposition was present at Creation. It had to be. The beings that flowed from Source had to have free will to be true individuals. Free will cannot exist without choice. Their choice was whether to align with God's will or pursue their own desires, that is, to go in an *opposite* direction from the Whole.

The force called opposition, like all fallen spirits, undergoes different manifestations or intensities of itself in different dimensions. Satan is one of the manifestations of opposition on a lower plane. Satan, in the earliest Jewish tradition, was not the horned, cloven-hoofed demon of evil. He was the angel of opposition. Like the perverse Loki in Norse mythology, he presented continual challenges and temptations for humans to overcome, and in doing so prodded them to grow.

. . . we realize that the Spirit of Perversity and Darkness is the spirit that, according to the Dead Sea Scrolls, indwells within each believer who must struggle to overcome it so that the Light will shine and he or she may thus begin the higher ascent through the unification of the internal aspects of man's soul with the externalization of the Divine Crowns of Creation.[1]

The lower the dimension, the more opposition took on characteristics we would call evil. But, at the highest level, it was the choice to exercise individual will apart from the will of Source. To possess true individuality, the Aeons had to have an alternative existence. Opposition provided the ability to have thoughts and experience an existence that was of their own making, not God's. Making that choice is synonymous with exercising free will.

Instead of conforming to an orbit around God's will, spirits could go off on a tangent of their own making. In figure 5, the central sphere represents Source. The circles around it represent the Aeons, spiritual beings that have chosen to remain in harmony with God's will. The arrow represents spirits who chose to break from God's will to construct their own experience. Chaos is the area or mindspace beyond the organizing force of God's will, where the random chance, or potential of free will, could play out in some new fluid form, i.e., some new individual experience.

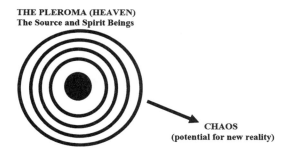

THE PLEROMA (HEAVEN)
The Source and Spirit Beings

CHAOS
(potential for new reality)

Figure 5. The First Spiritual Exercise of Free Will
(The Fall from Heaven)

The original meaning of the word *sin* is to be in error, or off the mark. If the Source and its heavenly surroundings of spiritual beings are the "bullseye," the arrow in figure 5 shows the original sin, or fall, as the tangent or vector that has indeed gone off the mark into chaos to create something new on its own. The story of Sophia is how spirit went off the mark.

THE BREAKING OF HEAVEN

Sophia's passion to know Source was causing her to press against the confines of the agreement that kept the Aeons within the Unity known as the Pleroma, or heaven. Barred from realizing the ultimate nature of Source, as all beings were, her mind began probing elsewhere to solve her dilemma. On the fringes of heaven, chaos existed like a shadow, an area of virgin potential left unorganized and undefined by the mind of God. It was a dimension of possibilities where something new could occur. In scientific terms, chaos is the quantum foam or zero-point energy field from which random realities can arise. I'll describe these phenomena in more detail in later chapters.

Authades, being the impulse of opposition, or self-will, caused Sophia to focus her attention on chaos. There, Sophia believed, she could create something in the manner that God created, and thus know the Source by tracking Its creative pattern.

Like all Aeons, Sophia had a male counterpart aspect. An Aeonic pair would consort between themselves, using the consenting power of Source, to produce a new Aeon pairing, or archetypal dimension, in a kind of closed-loop process. Sophia was the first to break this pattern of symmetry. Without the consent of either her male consort or the Source, she sought to create on her own, and broke with the Unity of heaven by plunging her essence into chaos.

Before we continue the intriguing story of Sophia, let's summarize and clarify the symbology of what you've read so far:

- God (Source) expressed the goodness of Its infinite Unity in the form of archetypal ideals of existence (truth, love, etc.).
- These ideals were represented by actual male-female pairs of spiritual beings (Aeons, or Eternities) with their own frequencies of thought energy that formed separate but connected dimensions collectively, known as Pleroma, the "Fullness," or heaven.
- The Aeons had free will (choice) to function as beings with a sense of individuality apart from the Source, though most chose

to use their will to remain in complete alignment with the will and intention of the Whole.

- The ability to use free will in a different, opposite direction from Source was necessarily present at the moment of the Creation. Without choice, there can be no real free will. This choice was embodied as the thought-form Authades (Audacity, Opposition).
- Authades influenced Sophia's thought of making her own unilateral creation. This represented the first exercise of free will. This was the original fall from the state of unified grace, where an individual being placed its will above that of the Source and acted for its own desires, ends, and means apart from the "hive mind" of the Unity.

Some of you may recognize the striking similarity here to the fall of Satan in Christian theology. Indeed, the Satan story most likely derived from a partial understanding of the Gnostic Sophia myth. I say partial because Gnostic cosmology was multidimensional, taking place through parallel universes or dimensions with more than one fall, as opposed to the singular biblical fall.

Gnosticism describes archetypal forces or beings, such as the angels and demons familiar to Judeo-Christianity, but these forces played out in increasingly corrupted multiple lower dimensions, away from Source. Orthodox Christianity, with concepts like Satan and a single fall, only perceived the tail-end lower effects of this dimensional fallout, like reading only the last chapter of a book. The Gnostic story relates events from their beginning, in the higher causal dimensions, all the way down to the lower planes of human existence.

Sophia was not evil, but her actions led to evil—some said as a cosmic error, others as part of a cosmic plan. But which was it? You decide, after reading on.

8

The Fall

How Spirit Descended

CHAOS

Infected by Authades (Audacity, opposition), Sophia plunged the essence of her divine mind and energy into the unformed potential of chaos. Here, she felt she might create as God created, and so know Source through Its creative power.

But chaos was not empty. As we know from modern science, vacuums don't really exist in nature, though on the surface there may appear to be voids. A substance called proto-matter was present in chaos, named from the Greek meaning "preceding matter." Quantum science would identify Gnostic proto-matter as dark matter, or virtual particles. Contact between Sophia's divine high-frequency energy and the proto-matter produced something quite startling and unforeseen.

The inert proto-matter activated upon contact with Sophia's high energy. The proto-matter particles swarmed around Sophia like iron filings flocking to a magnet, as depicted in figure 6. The interaction changed both forces. The inert, lower energy proto-matter was now moving and taking form, but Sophia's high-energy vibration was being lowered. Her light energy from the higher dimensions was "darkened," mutating into a new, lower form.

In the Gnostic text *Pistis Sophia,* Sophia cries out to heaven in alarm: "Save me out of the matter of this darkness . . . I am become as a material body . . . I needed my power, which they have taken from me. Save me quickly out of this chaos. . . . My time is vanished like a breath and I am become as heavy weighing matter."[1]

Figure 6. Sophia's Descent and the Formation of Matter

A NEW STATE OF BEING

Sophia's energy was morphing into something of a lower vibration and, mixing with the proto-matter, a denser substance like dark matter resulted—though not yet physical matter, which would come later. The new mindset or thought-form of individuality Sophia generated to produce this new dimension of matter in her lower energy state was opposed to the collective vibration of the Source. It took the shape of a new intelligence, an ego-centered being well known in the ancient world, the *Archon* (Gk: "Ruler") or *Demiurge* (Gk: "Craftsman").

This being was ignorant of the higher spiritual worlds above him, though he sensed the pattern of the higher realms. He began to manipulate his environment (matter), as the ego will do, into forming a universe that was a flawed shadow copy of the dimensions above him. This process eventually led to the formation of the material world, though it happened in stages. The first Gnostic fall described the creation of this psychic dimension of subtle matter, called dark matter today. Dark matter is invisible because it doesn't interact with light or other particles.

The appearance of this new thought-form dimension of mind and matter caused a great disturbance in heaven, according to Gnostic texts. Being a lower frequency, it could not exist in the same plane with high-vibrational spiritual energy. Thus, matter was "expelled in a great disturbance and projected apart." This ancient description is eerily reminiscent of the Big Bang.[2]

In the higher realms, *thoughts are things* according to both the ancients and many modern psychics. The thoughts of beings, or powerful centers of consciousness, in the higher planes can become organized, independent entities of their own. They could be conscious centers of energetic intelligence in varying degrees of permanence and form.

Sophia had the thought of separating from the Source to create on her own and, according to *The Apocryphon of John*, "her thought did not remain idle, and something came out of her which was imperfect." It was imperfect because she acted unilaterally *"without the consent of the Spirit, - and without her consort,"*[3] that is, without God (the Source) and the male aspect of her being. Her thoughts and actions ran counter to divine law, so the result was flawed and imbalanced.

Sophia's flawed thought-form of individuality was imbued with **Ego.** It took shape in the form the Gnostics called the Archon or Demiurge, a false mind-construct or "counterfeit spirit." This is the belief in separation instead of unity, the notion that we exist independently, in a duality separate from others and God, instead of recognizing we are all part of a Unity.

Sophia's actions generated the first conversion of energy into denser matter—the first instance of Einstein's $E=mc^2$. The texts mention vast quantities of matter being formed by Sophia's contact with proto-matter, but this was not yet the physical matter comprising the material universe. It was **dark matter**, a form of subtler, invisible matter that we'll discuss in the section on quantum physics. The texts are clear that this was not yet a physical creation but a kind of etheric dimension consisting of virtual particles, the precursor to physical matter.

Nonetheless, it was a very real new dimension of existence created and governed by a new mindset. It was the psychic or soul dimension where fallen spirits swam in a field of virtual particles, which they would use to fashion experiences according to their desires. This new mindset of consciousness governing the manipulation of matter was soul-mind or ego-mind, which is the use of God-given consciousness to feed the desires of Self.

So, the first creation of matter was born from a fallen consciousness where each soul acted independently to fashion creative experience according to its own ends. Naturally, the mindset governing this dimension of manipulated matter was egoic, because the ego promotes the desires and preservation of Self.

MORE ON THE ARCHON (DEMIURGE)

Religion and psychology recognize three states of being—spirit, soul, and matter; or spirit, mind, and body as commonly referred to today. Soul and mind have always been equated. In Greek, the soul is called *psyche,* indicative of the functions of the human mind. Sophia's actions produced the psychic or intermediate dimension of mind between spirit and physical matter.

The Archon (Demiurge), the mysterious being Sophia brought forth that governed the chaotic new matter, is symbolic of the ego-mind. In the Gnostic texts, the Archon generated other, minor Archons that represented a host of negative archetypes, such as greed and envy. The Archon represents the lower mind function of our being, which focuses on manipulation of the world around us for self-preservation and gratification. It's the focus of mind toward personal need and desire, as opposed to the consciousness of a higher unity.

The Archon, as pictured in figure 7 on page 56, was said to have the head of a lion with flashing eyes, and the body of a serpent. Okay, I know, that sounds like a sword and sorcery script—the Archon figure has made its way into the video gaming world—but it is important to understand the underlying symbolism of the Gnostic Archon imagery. In the ancient world, the lion was symbolic of fierce pride. The serpent was a symbol of wisdom—and remember, Sophia was the archetype of wisdom.

So, the Archon as a lion's head governing a serpent's body represents pride and ego usurping the higher judgment that comes from wisdom. It represents the illusion of individuality and ignorant dualism that is blind to the reality of Oneness—rash, chaotic, and self-gratifying impulses prevailing over reasoned, organized, harmonious thought.

But if Sophia was wisdom, how did she get herself into this mess? Well, how does one acquire wisdom? Mostly through trial-and-error

Figure 7. Sophia Births the Archon Ego-Mind (Demiurge)

experience, and Sophia was the pioneer who gained experience in a new creation apart from the Source. You don't truly experience how fire burns unless you stick your hand in it, and she did just that.

To show how this Gnostic allegory lasted through the ages, look at the medieval Fool tarot card in figure 8. It shows an oblivious wanderer with his or her head in the clouds, about to fall over a cliff.

The wandering fool is Sophia, the wisdom aspect of God that fell into the lower planes in an act that can be viewed as rash, oblivious foolishness. The Fool tarot is symbolic of spirit falling from a high state of unity into a lower plane of individual experience and suffering.

Sophia's fall produced a new ego-mind state, a psychic dimension of soul, which is spirit seeking personal experience. The qualities attributed by the Gnostics to the Archon scream ego attributes—prideful, imperfect, petty, jealous, limited, weak, self-centered, envious, and ignorant of higher consciousness or the existence of a higher order outside itself.

*Figure 8. The Fool
from the Classic
Ryder Tarot Deck*

Let's summarize:

* Sophia's actions departing from Source were the result of free will choosing a separate path of pursuing individuality as opposed to harmonization with the Whole.
* This thought-form of individuality apart from the unified consciousness of Source became a new, lower universe or psychic dimension of individual minds or souls pursuing personal creative experience.
* The Archon (Demiurge) represented the rise of an ego-minded consciousness necessary for pursuing selfish individual desires and self-satisfaction.
* This ego-mind would go on to create material experience in a deluded vacuum, thinking its reality was *the* reality, ignorant of the superior forces that preceded it.

That last point describes the origin of the modern atheistic-materialist ego-mindset of many people today. Having no awareness of God (Source), the existentialist view is that we are solitary creatures with no higher purpose. We're ignorant of how we arrived here and where we are headed.

In the next chapter, we'll see how this psychic, ego-minded force of fallen souls formed the material world as we know it.

9

The Force Of Opposition

The Dark Side of Spirit

To understand the meaning of the Archon, we need to revisit the universal force of opposition I alluded to earlier. It's a topic that needs elaboration, as it helps to explain the tragic history of the suppression of mystical wisdom as well as the ordinary friction or opposition we experience in life. Opposition, or polarization, permeates the universe but, since science and psychology have no real answers about from where this phenomenon arises, we'll look for spiritual explanations instead.

GRASPING OPPOSITION

I've described how the original spiritual engine of Creation used the energy between polar forces to power the movement of Creation, but no opposition existed at that level; the polarities were in harmony. Opposition in the higher dimensions existed in *potential* as a necessary force to enable real free choice, which in turn was necessary for individuality to exist distinct from the Source.

Individual spirits at first used their free will to stay in alignment with Source. Opposition came into play only after spirits exercised their choice to pursue their own individual paths, in other words, to go in opposite directions from Source's will. Opposition at this level was not evil but it led to evil in the lower dimensions as spirits (now called souls) deviated further and further from the Source with increasingly selfish desires.

Opposition can take many forms. It can manifest as obstacles that stand between you and your desires. It can take the form of internal conflicts between the heart and mind, or between the conscious and unconscious minds. It can be antagonistic people that seem to pop up in every walk of life. For every Hitler, a Churchill arises; for every ideology, a counter-ideology arises. We speak in this book about materialism in opposition to spirituality or consciousness.

A deeper dimension exists to opposition, however. As I've said, opposition was a necessary element to allow individuality to come into being by presenting an opposing choice for the exercise of free will. But once the choice was made to lower consciousness in order to experience the illusion of material existence, the role of opposition flipped. Opposition became an obstacle to be overcome.

Both science and the Gnostic texts indicate that we're living in a simulation, largely unaware of an underlying reality. This stands in opposition to our central life's purpose to awaken to higher consciousness by connecting with that part of us that *is* our higher consciousness. Opposition, which in the higher planes promoted the opportunity to break with the Source and go our own way, now assumes the role in the lower dimensions of keeping us from our central purpose of working our way back up again.

The Gnostic Archon is symbolic of the collective ego-mind or the force opposing realization of the Source. But the Archon is a kind of pseudo-life-form, an artificial construct, just as this world is an artificial reality. The Gnostic texts say human souls are still spirit, but the Archon is an energy thought-form devoid of spirit. It owes its existence to the illusion of a separated existence from the Source—it dies if everyone exits the hologram back to where we came from. Indeed, that's exactly what the Gnostic texts say the ultimate fate of the Archon will be—dissolution.

So, the force of opposition is given life by our own mistaken belief in duality instead of unity. The belief that we are separate threads and not part of a whole tapestry bars our way back to the higher dimensions of consciousness. It works through our minds planting lies, deceits, illusions, and false ideas while fostering limiting doctrines and control mechanisms such as cults, religions, government intrusion, and many other means. It's like an internal civil war where one part of us, locked away in the shadow of the unconscious, works against another part of us that seeks to elevate our existence.

Getting back to the Gnostic Sophia myth—Sophia activated a new mindset when her divine energy plunged into unformed chaos and the energy field responded, resulting in a new dimension. This was the psychic realm, the abode of ego-mind or soul-mind, for *psyche* means "soul" in Greek. It was the source or precursor to the physical dimension.

This dimension contained all the elements from which material reality would be fashioned. It contained virtual particles, subtle matter, and a new egoic consciousness that manipulated it. The subtle realm would contain elements like dark matter and dark energy, the subatomic precursors to visible matter. The new entity, the Archon, is representative of a new focus of mind oriented toward exercising its own brand of Creation. That's what ego does—it operates in ignorance of and separated from a higher pattern.

THE DILEMMA OF THE ARCHON

The Gnostic Archon raises a somewhat complex question. We've equated it with the rise of a collective ego-mindset, the new mindset that would effectuate creation at lower vibratory levels of existence using the newly formed subatomic particles. The description of the Archon in Gnostic texts certainly reinforces this egoic image. The Archon is arrogant, jealous, petty, territorial, and ignorant of higher forces despite his seemingly great powers.

Because of these shared qualities, many Gnostics equated the lower creator Archon with Jehovah, the tribal Jewish god of the Old Testament, and they contrasted him to the God of love, the higher revelation brought by Jesus.

The question that arises is, what is the exact nature of this Archon? Is it a real entity or just allegorical? The answer, and this is inferred in the Gnostic texts, is both. Surely, the characters of the Gnostic texts, like Sophia and the Archon, are not invisible people, but that is not to say they are real entities. The best way to describe them is like force fields or organized currents of intelligent energy, each focused on specific aspects that organize and maintain Creation.

To our human minds, which both receive and perceive beyond the five senses, these forces translate as archetypes—thoughts, images, and feelings that impart qualities and meaning to the array of energetic, electromagnetic manifestations we view as objects. They also manifest as the thoughts and feelings we experience as motivational directions, either positive or negative.

The Archon, however, is a bit of a special case. I've stated that the Archon is the collective product of fallen soul-minds operating at the psychic level, the vibration of egoic soul-mind reinforced by the souls fallen into the new dimension. These newly minted souls, though in a fallen state, still retained great power.

The axiom "thoughts are things" becomes truer the more you climb the dimensional ladder. Indeed, the Gnostic texts repeat that "thoughts are not idle." The soul's desire for individuality gave rise to a collective ego-mind or false belief in separation, which the Gnostics refer to as the counterfeit spirit.

Souls seeking self-satisfaction and individuality poured their thought energy into this image and idea until it assumed a transpersonal reality as the Archon. This force of opposition appearing everywhere in every life is so pervasive, we must assume it has a source, an organized intelligent center from which it radiates, first through the mind, then becoming externalized as experienced events. This indicates a psychic—Gnostics equated psychic, soul, and mind—rather than a physical origin.

The proof of the Archon's psychic quasi-existence is in the Gnostic texts. It is described as having soul, but not spirit. All souls possess spirit originating from Source, and humans are embodied souls. The unique soul-not-spirit status identifies the Archon as being an entity purely arising from the psychic (soul-mind) dimension. The irony is that the Archon thought-form was generated from individual soul-mind energies but then proceeded to rule over them collectively. Souls, then, are of a higher, more permanent order than the false idea into which they've invested.

This paradox harkens to a passage in the Gnostic Gospel of Philip, which says "... *men create God. That is the way it is in the world—men make gods and worship their creation.*"[1] Another way to state this is that we suffer as slaves of our own errant thought patterns, something with which most modern psychologists would likely agree.

This is where illusion comes in. You could say that we ran a software program that created a virtual reality necessary to delude ourselves into believing we are separate physical beings in a world of solid objects. Now we're prisoners believing in our own program. The Archon is like Agent Smith in the *Matrix* movies. Smith was a program that went wild and assumed a type of reality that came to control and affect the very computer network he was created to serve.

Once established, the ego force, the program or opposition to higher awareness, must do everything it can to stay viable, and that means keeping human minds in bondage to its simulated reality. We can say that the Archon force has assumed a *psychoid* existence, to use Jung's term (psychic with physical manifestations), fed by our ignorant thoughts and negative emotions.

Psychoid is a vital principle directing the behavior of an organism, in this case the principle of opposition. As Jung said, the psychoid is the bridge between psyche and matter. I believe the Archon force, the principle of resistance and opposition, resides upon this bridge. We may have created this program at a psychic level of existence but as long as we subconsciously or unconsciously accept it as real in our thoughts, actions, and emotions (and we all do), it is real. As the Gospel of Philip alluded, we've become prisoners of the God we've created. Indeed, the Archon is pictured as a parasite promoting and feeding off our fears and misery.

In the Gnostic texts, the Archons are pictured residing in the firmament, the inorganic wider cosmos or universe distinct from the living organic Earth. They're also tied to the concept of deterministic fate. This makes sense since they're devoid of spirit, which imparts purpose, intention, and free will. A deterministic existence is one seemingly governed by fate and caused by mechanistic, materialistic laws such as Newtonian mechanics. The inorganic matter of the universe apart from Earth behaves in this mechanical way. The Earth, with its conscious bio-spiritual life-forms, is different.

In Gnostic terms, we contain a spark of the Master Consciousness of Creation through the wisdom of Sophia, the Earth Mother. As we'll see in later chapters, quantum science reflects this creative power in our ability to determine particle outcomes by observation (double slit experiment). Reality is dependent on us as observers, but we are not dependent on random events of the simulation (fate) unless we believe it to be so, which, unfortunately, is where materialism directs us.

The sole purpose of the principle of opposition on this plane, then, is to obscure our vision and prevent us from recognizing our true origin and power (the Source). It does this through false beliefs and philosophies such as materialism, which tell us we are random occurrences of particles subject to random events of chance and fate.

In the Gnostic texts, the Archon is pictured as a gatekeeper who, unless a soul has thrown off the shackles of illusion, will repeatedly throw the soul back into reincarnated lives bound to the Earth plane, thereby perpetuating its own power. In reality, our own ignorance is our true gatekeeper.

An interesting allegory in the Gnostic gospels describes how the Archon tries to rape the biblical Eve, but fails.

"What sort of creature is this luminous woman? . . . Now come, let us lay hold of her and cast our seed into her, that she may become soiled and unable to access her inner light. Then those who she bears will be under our charge" . . . But Eve, being a free power, laughed at their decision. She put mist in their eyes [and escaped them].[2]

This is a richly symbolic and highly informative passage. It describes Eve as luminous, meaning endowed with higher consciousness. The Archon's goal is to enslave superiorly endowed humans by severing them from their higher consciousness, making Eve *unable to access her inner light.* The rape is the attempt by the Archon (the force of opposition) to cause humans to *be under our charge,* infecting them with its mind parasites of false ideas, to force humanity into a defiled, lower state of illusion. The failure to consummate with Eve, who has *"free power (free will not subject to fate),"* means the illusions of the material world are ultimately doomed to fail at permanently enslaving the soul as humanity awakens to its true, higher spiritual state. The central message of

Gnosticism is about transcending the lower material world.

I used the term "mind parasites" because harmful ideas can infect anyone like a virus and keep them in a state of delusion without the victim knowing it's infected. Just as parasitic viruses subvert the body's functions, so do mind parasites pervert the mind's functions. Nazism and communism can be likened to ideological viruses or parasites that enter the mind, subvert a person's normal behavior, and cause them to do great harm, thinking they're doing good. This is true for any cult. It was true for early Christians, like the ones who killed Hypatia, and it's true for present-day Islamic radicals. The texts describe this as a *blind chain of compulsion.*

I'm not the first to make this connection. Carlos Casteneda, in his famous books about Don Juan the Mexican sorcerer, describes shadowy figures on the edge of our perception, which he calls flyers, that are eerily similar to the Archon(s), or mind parasites. (I use plurals because the Gnostic texts describe the chief Archon spawning others, representing negative impulses like hate, jealousy, and envy.)

Castaneda says these flyers came from the depths of the cosmos to enslave humanity by implanting us with false, negative ideas and belief systems. He says the flyers give us our minds, which is *their* minds. Whether or not you believe in the overall veracity of Castaneda's work, he describes a phenomenon completely analogous to the *counterfeit spirit* in the Gnostic texts. This is the effect of the force of opposition that makes us mistake and accept the illusory, suffering nature of the world for reality.

Once we raise our consciousness, we can detect this unconscious parasitic activity in our minds and take steps to counter it by connecting with that within us that is infinite and eternal as opposed to what is finite, illusory, and misleading. This was the promise of the Gnostic masters and the Mystery Schools. Understanding the play of the forces I've described here will enable you to better grasp the profound insights of the Gnostic masters.

10

The Truth of the Matter

The First Appearance of Matter

BIBLICAL SHORTCUTS

Though Gnosticism predated Christianity, Gnostics were among the first Christians. The Judeo-Christian biblical Creation story was likely drawn from the same roots of Jewish-Kabbalistic mysticism already present in Gnosticism, but the orthodox Christian version picked up only the tail-end of the Gnostic story.

In the Gnostic view, many orthodox Judeo-Christian beliefs were dumbed-down, misinterpreted versions of more complex events. They felt that orthodox Christianity heavily edited and trimmed the deeper mysteries for mass consumption. In cinematic terms, they left the most critical material on the cutting room floor.

For example, the Gnostics essentially described three Creations and two falls. God created heaven, the harmonious Aeonic dimensions. Sophia then broke that unity, and her fall in consciousness created the psychic dimension of mind and soul. From that dimension, a *second fall* occurred, resulting in the material Creation. A biblical hint of this graduated Gnostic progression leading to the material creation does exist if one knows how to interpret the apparent contradictions between Genesis books 1 and 2. However, the way things were taught, God created the world and two human beings fell from grace in an Earthly garden. The concept of intermediate dimensions and beings (still recognized in Hinduism) was lost in the process, much to the detriment of our ability to grasp reality, as we shall see.

LOST SOULS

When the dust settled after Sophia's fall, two major dimensions of being now existed—Heaven (the Pleroma), the world of Higher Mind and Consciousness focused on unity and harmony; and the psychic plane of lower mind or fallen spirits (now called souls), which focused on individual pursuits.

For clarity, I want to draw a distinction here between spirit and soul. Both have qualities of individuality. However, *spirits* remain in a higher vibration, using their individual will to remain in harmony with the Source. In biblical terms, they remain before the throne of God. *Souls,* by contrast, are spirits pursuing individual experience in lower planes of existence. The record of this experience is retained by the soul-mind over lifetimes of physical incarnations and nonphysical states in between, according to mystical belief. *Spirits are not souls, but souls are spirits in a lower state of consciousness.*

The psychic plane was chaotic and focused on individuality. Why was it chaotic?

+ Numerous souls with often conflicting desires. . . .
+ No longer able to see the guiding light of Source. . . .
+ Experimenting in a vacuum with new, untested forms of being—enough said.

The soul entities lured into Sophia's accidental Creation were now *really* far from Source, each getting lost in their separate creative dreams under the Archon, ego-minded influence. They still had glimmers and yearnings of their former blissful existence but their baser conflicting desires short-circuited the higher impulses that beckoned them.

Like in popular post-apocalyptic movies where civilization is destroyed, the survivors have to figure out how to fend for themselves with what's left. They have memories of better, civilized times on one hand, yet a wide-open new world to create on the other. What to do now, which way to turn? The souls' choice led them into material existence, a condition even further afield from their origins.

THE MATERIAL WORLD

It's important to understand that in the Gnostic belief system and that of mysticism in general, all dimensions, thoughts, and creations are based on the original pattern from the mind of Source. The lower psychic and material planes imitated this pattern but with significant distortions.

Just as Source manifested spiritual dimensions of Aeons, the imitating Archon force manifested a series of baser dimensions and beings, or negative archetypes, each with its own lower level of consciousness, given names like *envy* or *hate*. These qualities were reflected in mystic lore as being attached to the astral and etheric planes.

Patterns for the creation of the universe, the Earth, and the human being had always existed in the mind of Source. They filtered down through the dimensions in an ethereal template, but had not yet become physical. We know from the Gnostic texts that matter preexisted in finer forms in the nonphysical dimensions. The Archon force—think of it as the collective ego-mind of the fallen souls—fashioned this subtle matter to create the psychic dimension. This same force lowered the vibration of the subtle matter to take on mass and fashion the material world in imitation of the higher pattern above, but with many flaws and deformations. An allegorical image may help us picture these dimensional distortions.

Say a pure lake exists high atop a mountain. As water from the lake flows downhill, it picks up various contaminants. This stream originates from clean water, but in its downward journey becomes increasingly polluted while the source at the top remains pure. Visualize that, about halfway down, we acquire pollution at the psychic level, then down at the bottom we collect more pollution at the material level. This, incidentally, is the Gnostic answer to the problem of *theodicy,* the paradox of how an evil world can come from a good God.

God, the Supreme Being, did not create this world. The idea that the world was shaped by lower beings variously called Archons, angels, souls, or "gods," was not limited to the Gnostics. Jewish sects, such as the Magharians, held that the world was created by angels. Papias, an early Christian apostolic father, said, *"To some of them [angels] He gave dominion over the arrangement of the world, but the arrangement failed."*[1]

So it was that soul-mind forces of the psychic dimension, inferior to their Source, created a new material playing field to further lose themselves in their dream creations. **They projected their consciousness into the three-dimensional dream of materiality.** The distortions of their diminished consciousness resulted in a world of light and shadow, beauty and suffering, good and evil.

It was called Earth.

Sophia, seeing the chain of events she'd set in motion, repented. Her penance was to remain in the lower dimensions as a ray of light to counter the archontic influence, which perpetuated its existence by feeding off the dualistic illusions of fallen souls. So it is that the spark of the divine remains within us, struggling to break out of our false beliefs and return to our Source.

The ego-driven force of the Archon, like our personal egos, ever seeks control, wanting to stay in the driver's seat. Since the psychic realm was a kind of contested dimension between light and shadow, what better way to control the fallen soul-minds than to drive them further away from Source and into material existence? The Archon perpetuates its artificial existence by sowing the human mind with false beliefs, or mind parasites, to create the fear upon which it feeds.

Since a pattern, or blueprint, for the material creation was already present throughout all the dimensions, a "psychic" Earth already existed in thought-form. This accounts for noted differences in Genesis books 1 and 2. Book 1 describes events on this psychic "pre-Earth," and book 2, on the physical Earth.

The Archon force usurped and twisted this Creational blueprint into a material dream world. Using psychic energy, it fashioned the material cosmos and physical life-forms into which the fallen souls poured. The Archon force became a false god, usurping Source by fashioning a counterfeit reality of dualism and separation instead of the true condition of a singular unity. Of course, this Archon was a collective illusion we created, then allowed to master us.

The Source, though allowing parts of Its consciousness to undergo these experiences, would not leave souls to be lost in a perpetual dream—that would be divine insanity. Despite all the chaos and distortion, a spark of the divine still remained in each fallen soul. So, It created a safety valve in the form of the repentant Sophia working with the other spirits from the higher planes. This force counters the grip of the Archon ego-mind on the human soul and seeks to awaken that divine spark.

Hence, the principle of a saving grace or awakening to the light originated to stir the ability in each of us to return to the wholeness and bliss of our higher consciousness. So began the seemingly eternal struggle between light and shadow. Now we'll see how that conflict manifests in the lower dimension we call our Earthly home.

11

Jewish Mysticism

Moses and the Origin of the Kabbalah

The link between Jesus and Gnosticism we will discuss in the next chapter must be understood in the context of Jewish Mysticism, and that story begins with Moses. The enlightenment Moses received at the Sinai came in two forms, written and oral transmittals. The written law was the *Torah Shebichtav,* and the oral tradition was *Torah Shebaal Peh*. The oral tradition was supposed to descend from wisdom inherent in Adam and passed down to Abraham.

Figure 9. Moses and the Commandments

In an article by Rabbi Nissan Dovid Dubov, on Moses, Rabbi Dubov describes the mystical meaning of the Sinai revelations as follows:

> Kabbalistically, the revelation at Sinai was the breakage of a barrier. In the words of the Midrash, prior to Sinai, it was decreed that the upper realms could not descend below, and the lower realms could not ascend above. At Sinai, the upper realms descended below and the lower realms ascended above.

The meaning of this cryptic Midrash is the following: prior to Sinai, there was a divide between the body and the soul, between the physical and the spiritual, between Earth and heaven. It was not possible to endow material objects with innate spirituality. The spiritual elite were endowed with powerful souls and minds that were fully aware of the mystical dimension, and yet had not integrated that system in a down-to-earth practice for all mankind. Sinai changed everything.

> **God revealed to Moses his plan for the fusion of heaven with the Earth and the material with the spiritual** [Bolding mine]. From then on the barrier between these two seemingly opposed realms were not to be two separate entities but fused into one, and once again, the Shechinah [Sophia] dwelled amongst Men.[1]

Early on, I independently intuited humanity's role as the bridge between an ultimate merging of the higher and lower dimensions. I said in my lectures and previous books that **the purpose of humanity is to spiritualize the material and bring the experience of material back to spirit so that all planes of existence align in harmony and all separation dissolves for the illusion it is.** I emphasized Rabbi Dubov's sentence above because it validates this insight.

I will later cite a passage from the Gnostic gospel of Jeu, which concurs on this exact revelation. This is no coincidence. Jewish mysticism and Gnostic beliefs from the East had largely cross-pollinated. Like my vision, they were based on insights gleaned from higher dimensions. In the Kabbala, Shechinah and Hokhma are the feminine figures representing God's presence in the world and God's wisdom respectively. Shechinah, the mother of humanity, was associated with the Holy Spirit. The feminine Sophia in Gnosticism was all these things combined, so I added Sophia (in brackets) to Rabbi Dubov's quote above.

THE ESSENES

The Essenes, who could be called nonconforming Jews, lived in scattered communities all around ancient Judea. They represented one of the three major Jewish sects, along with the Pharisees and Sadducees. Ancient writers, such as Pliny and Philo, describe the Essenes as originating far back in antiquity, even unto the time of Moses. Church father Eusebius quotes the great Jewish scholar and biblical interpreter Philo of Alexandria, who calls the Essenes "the holy ones," and says that:

> [T]en thousand of them had been initiated by Moses into the mysteries of the sect. Our lawgiver, Moses, has trained thousands of disciples who, on account of their saintliness, I believe, are honored with the name of Essæi.[2]

Rabbi Harvey Falk, in his book *Another Look at the Jewishness of Jesus,* confirms that the Essenes descended from ten thousand people trained in the mysteries by Moses. When Moses brought the first commandments down from the mountain and saw the fallen state of the Jews, he broke them, realizing they were not ready to receive the divine revelations. He later gave the masses a second set of commandments, but the Essene mystics he trained from the first set of unrevealed tablets.

The Essenes were the mystics and healers among the Jews, and were often at odds with the other dogmatic Jewish sects. Various speculations on the origin of the name "Essene" are: "modest," "humble or pious ones," "silent;" or, from the Aramaic "asa (to heal)," or "the healers;" from "'asah (to do)," with reference to the "'anshe ma'aseh (the men of wondrous practice)." Whatever the derivations, they're all consistent enough for you to get the general idea of what they were about.

Though scholars have questioned some of his writings, philologist

Figure 10. Essenes

Edmund Szekely says the esoteric Essene teachings were part of the universal ancient (perennial) tradition, and were described in several writings, most notable being *The Tree of Life.* He correctly points to the common threads of the teaching, as found in Persian Zoroastrianism, Indian Hinduism, and Tibetan Buddhism. Gnosticism stems from the same sources.

The Essenes were, according to biblical-era historian Josephus, regarded by King Herod as endowed with higher powers (Josephus, *Antiquities.* xv. 10, §§ 4-5). One Essene, Menahem, possessed the divine gift of prophecy. Josephus relates that he once sat in the Temple surrounded by his disciples, whom he initiated into the (apocalyptic) art of foretelling the future. (*Ant.* xiii. 11, § 2; B. J. i. 3, § 5).

Jewish historian Flavius Josephus tells us that communities called Essene were dispersed all about Syria and Palestine in biblical times. Nearly all ancient sources agree on this point.

Before Jesus, many communities existed around Judea that were not in direct obedience to or agreement with the High Priests and the Jerusalem Temple.

These groups had differing practices, but they did have some things in common: they were separatists; they lived communally; they performed cleansing rituals; they often wore white robes; they were versed in the healing arts; and they were mostly, if not exclusively, vegetarian. They also believed in the coming of a Messiah, though their definitions of that concept may have differed. For the most part, they rejected the Temple priests of that era as usurpers of the ancient priestly line of Zadok, a change that occurred during the time of the Maccabees.

Essenes have been popularly associated in recent times with the group of ascetic purists that wrote the Dead Sea Scrolls, which were found inside the caves of Qumran. This was a strict group of largely puritanical men preparing for Armageddon. They were an extreme branch of what we broadly called Essenes, and a likely source for the Zealots and the Sicarii who assassinated Romans and collaborating Jews.

> For some of these observe a still more rigid practice. . . . Others again threaten to slay any Gentile taking part in a discourse about God and His Law if he refuses to be circumcised from this they were called 'Zealots' [Kanna'im] by some, 'Sicarii' by others. Others again will call no one lord except God, even though they be tortured or killed.[3]

This quote by church father Hippolytus hints at the extreme differences among groups broadly called Essenes. Some scholars feel that John the Baptist may have come from this Zealot group, pointing out that its teachings were substantially antithetical to Jesus. This apocalyptic group was not the same as the Essenes from which Jesus likely came, and who, among other things, were allowed to marry, have families, and were peaceful.

Mount Carmel was home to this latter group of Essenes. The Carmel headland is a promontory jutting out from a mountain range in northern Israel, which stretches southeast from the Mediterranean Sea. It has been considered a holy place throughout history, at least early on by the Egyptians, since the fifteenth century BCE. The biblical book of Kings describes an altar to God there, and Carmel was the

Figure 11. Mount Carmel

retreat of the great Hebrew prophet Elijah; his cave is still displayed there, to this day.

We also have strong evidence of the presence of one of the ancient mystery schools atop Carmel. The historian Tacitus says an oracle was present on the mount with whom Emperor Vespasian once consulted. Carmel was the dwelling place of the great prophet Elijah, and his temple once stood there. Other fascinating historical evidence indicates that Carmel was a hotbed of mystical activity.

The ancient presence of a mystical school on Mount Carmel teaching a Jewish variety of the universal or perennial wisdom tradition, is further bolstered by the biography of Pythagoras, the famous mathematician-philosopher and founder of one of the greatest mystery schools in the Mediterranean, at Crotona in Greek Italy.

Iamblichus, in *The Life of Pythagoras,* says that Carmel was *"more sacred than other mountains, and quite inaccessible to the vulgar."*[4] This statement would be understood by any ancient mystery school acolyte from India to the British Isles. The uninitiated, or spiritually immature, were not permitted up the sacred slopes.

Carmel means "garden paradise" in Hebrew. In the Judaic tradition, the garden of paradise was the place where humans reached their full potential under divine law. This is one indication that holy Carmel was a great spiritual center long before the birth of Jesus, and so, Pythagoras made a pilgrimage there.

And what exactly was the great Greek philosopher-mystic doing on

Carmel? He was on a spiritual odyssey, being trained in all the mystery schools around the Mediterranean. Iamblichus tells us the young Pythagoras was coming down from a temple on the summit of Carmel when he encountered some Egyptian seafarers who:

> . . . began to reflect that there was something supernatural in the youth's modesty, and in the manner in which he had unexpectedly appeared to them on their landing, when from the summit of Mount Carmel, which they know to be more sacred than other mountains, and quite inaccessible to the vulgar, he had leisurely descended without looking back, avoiding all delay from precipices or difficult rocks.[5]

So, centuries before Jesus, a great mystery school was already well established on Carmel. Legends say that Jewish mysteries, such as the Ma'aseh Bereshit and the Ma'aseh Merkabah rites, had been taught to initiates upon that very Mount for generations, since the time of Moses and Elijah.

According to church father Epiphanius and the Jewish historian Josephus, Mount Carmel had been the stronghold of the Essenes who came from a place in Galilee named *Nazareth*—coincidentally, the hometown of Jesus. In antiquity, Carmel was considered so holy, they allowed no permanent structures on most of the Mount. It was dotted with white tents used temporarily in the warmer months; during harsher weather, the people broke camp and stayed in sturdier dwellings in towns like Nazareth, which surrounded the escarpment. Sources like Philo and Josephus speak very highly of these gentle people for whom healing was a central principle.

Jesus's affiliation with the Essenes could have extended beyond Carmel. Another well-known community of Essenes, called the Therapeutae, lived in Egypt by Lake Mareotis, near Alexandria. Jesus's family lived in Egypt for some time, as the Bible records. So it stands to reason that the family of strangers in a strange land would have sought refuge with a colony of their own people.

Many scholars certainly believe that Jesus and his family belonged to the Carmel Essene sect and that Jesus was highly influenced by them, at the very least. No smoking gun exists telling us directly that Jesus was an Essene, but the circumstantial evidence is overwhelming:

- He lived in Nazareth in the shadow of Mount Carmel, a mystical stronghold and a town loaded with Essenes.
- His radical Essene-like teachings and healings directly opposed the Pharisees and Sadducees.
- He was, as we'll discuss, a radical feminist, while the other patriarchal sects put women down. The Essenes were far more accepting of women, and their philosophy leaned strongly toward the feminine principles I've mentioned.
- When the young Jesus and his family fled to Egypt, as strangers in a foreign land they would likely seek a familiar culture, such as the Therapeutae Essenes.
- The Therapeutae forsook worldly goods for higher devotions, as Jesus bade his disciples do.
- Essenes, like later Christians, had a rite of immersion, or baptism, associated with healing body and spirit.
- The Essenes zealously guarded their sacred teachings, as the Bible tells us Jesus did when he would only speak parables to the masses but taught his disciples the mysteries in secret.
- The Teacher of Righteousness, a central character in the Essene Dead Sea Scrolls

. . . would seem to be a prototype of Jesus, for both spoke of the New Covenant; they preached a similar gospel; each was regarded as a Savior or Redeemer; and each was condemned and put to death by reactionary factions.[6]

Let's examine some other connections from ancient records that lead to the conclusion that Jesus taught a secret Gnostic-Mystic tradition.

The name he bears, Jesus the Nazorean, has northern sectarian implications. . . . The name borne by the earliest followers of Jesus was not Christians: they were called Nazoreans [Nazarenes] . . . The old Nazareans [Nazarenes], like the Samaritans, were opposed to the Judean traditions, holding that the southerners had falsified the Law of Moses.[7]

For the view has become tenaciously held among them [Nazorean Essenes] that whereas our bodies are perishable and their matter impermanent, our souls endure forever, deathless: they get entangled, having emanated from the most refined ether, as if drawn down by a certain charm into the prisons that are bodies.[8]

The quotes above tie the Essenes squarely to Gnostic conceptions of the soul upon birth and death.

There are also among them those who profess to foretell what is to come, being thoroughly trained in holy books, various purifications, and concise sayings of prophets. Rarely if ever do they fail in their predictions.[9]

To those who are eager for their school, the entry-way is not a direct one, but they prescribe a regimen for the person who remains outside for a year.[10]

The initiation practice in that last quote is a hallmark of the Gnostic Mystery Schools. That Jesus was of this Nazorean mystical tradition is indicated in the Gnostic Gospel of Philip, and in the biblical book of Mark.

The apostles who were before us had these names for him:
"Jesus, the Nazorean, Messiah . . ."[11]
He shall be called a Nazarene.[12]

This is convincing evidence that Jesus came from and most likely taught a Gnostic-Kabbalistic tradition long held amongst the community of Carmel Essenes. Understanding this background is important to grasping how Christian Gnosticism evolved from earlier Gnostic and mystical traditions.

12

Gnostic Jesus?

The Secret Teachings of Jesus

'm going to digress in the next chapter and give you a short fictional account of what it might have been like to have lived in an intellectual hotbed of the Mediterranean during the time of the Christ. I do this to underscore the spiritual ideas outlined in the preceding chapters, and also to demonstrate how the force of opposition manifests in practical terms in the material world. It is the story of how these forces destroyed Gnosticism, pushed female leaders out of the early Christian movement, and suppressed the mystical teachings of Jesus.

However, before our imaginary tale, some actual historical background:

THE HIDDEN MYSTICAL TRADITION OF JESUS

Jesus said to the disciples:

> Unto you it is given to know the mystery of the kingdom of God;
> But unto them that are without, these things are done in parables.[1]

This same passage appears in all three synoptic gospels. Jesus is saying that the disciples were among those receiving special mystery teachings while the public at large received parables tailored for mass consumption.

Noted early church fathers Clement and Origen also affirm the existence of secret mystical teachings within the body of the early Christian community. Clement says that, after the execution of Saint Mark's mentor, Saint Peter, Saint Mark came to Alexandria from Rome. Mark composed a secret gospel for those who were "being perfected." But, Clement goes on to say,

> Nevertheless, he yet did not divulge the things not to be uttered, nor did he write down the hierophantic [mystical] teaching of the Lord.[2]

By not committing to writing the most sacred teachings, Mark is following the ancient mystery school prohibition against transmitting those teachings in anything but oral form. In a letter called Secret Mark, Clement says that even the written version of Mark's secret gospel:

> is most carefully guarded, being read-only to those who are being initiated into the great mysteries.[3]

The brilliant church father Origen echoes Clement, saying that Jesus privately taught the disciples mysteries that were not preserved because they were too deep:

> to be conveyed to the multitude in writing.[4]

BELIEFS OF THE FIRST CHRISTIANS

Apart from the Bible, we have the Gnostic gospels showing Jesus's explicit Gnostic teachings. The Gnostics, who predated Jesus, recognized the content of his mystical teachings and became among the first, if not the very first, Christians.

The Gnostics saw Jesus as the highest expression of the expected illuminators. The illuminators were the prophets predicted to incarnate from age to age to dispense sacred wisdom and put humanity back on track when it was in danger of sliding into complete ignorance of its spiritual origins.

This ancient belief in divine appearances is revealed in the *Bhagavad Gita,* part of the great Hindu epic *The Mahabarata,* where the deity says:

When Righteousness Declines . . . when Wickedness Is strong, I
incarnate from age to age, and take Visible shape, and move among
men, Supporting the good, thrusting back evil, And setting Virtue
on her seat again.[5]

Jesus's teachings on the Gnostic gospels clearly derive from a different perspective than orthodox Christianity. For one thing, the focus
is more often on the *Christ spirit* incarnated in Jesus, which operates
from nonmaterial realms rather than the physical person of a carpenter
from Nazareth. The Gnostic gospels do not focus on historical or literal assertions about the Christian story. The nativity, the crucifixion,
and the resurrection, for example, are examined more for their symbolic
than their literal meaning. The Gnostic emphasis was to extract the
allegorical message from the events comprising the Christian story. In
doing so, they revealed universal information priceless to all humanity.

Many scientists and philosophers have noted how incredibly narrow
the window was that allowed life to come into being, from a scientific
standpoint. If the rise of life did indeed pass through the eye of a nee-
dle, it can be said of this world that ignorance and evil grow like a weed,
but enlightened wisdom is a delicate garden most carefully tended.

That said, let's take a brief journey into antiquity, to a critical time
in human spiritual history.

13

A Gnostic Tale

Story of an Adept in Ancient Egypt

Let me weave a story to help you imagine how life might have been in the intellectual melting pot of religions, mysticism, and philosophy that was ancient Alexandria, Egypt.

> **9 CE** - *Dion, a young man, walks the main street of Alexandria, Egypt, eating a flatbread and some cheese. Alexandria, a Greek city founded in northern Egypt, is a cosmopolitan center of learning, radiating Hellenistic culture throughout the Mediterranean.*
>
> *But the religious, philosophical, and intellectual currents of the*

Figure 12. Dion

city are by no means exclusively Greek. Egyptian, Jewish, Syrian, Persian, and other influences have contributed to a dynamic melting pot of cultures.

Dion is from a family of mixed Greek/Jewish parentage not uncommon in the wake of Alexander's conquest of Egypt and Asia Minor. Dion does not formally follow the Jewish religion, or the Law as it is called, but he is well aware of it.

Being part-Greek in a heavily Hellenized region, he is fascinated by the currents of Greek philosophical teachings that permeate the known world from India to Gaul (France), owing to the Greco-Roman Empire, which lasted from 350 BCE to the fall of the Eastern Roman Empire at Constantinople, in 1453 CE.

First Awakening

Dion walks the streets of the city, passing the busy merchant stalls and inhaling the smoky smell of roasting meat. He hears myriad languages being spoken. Groups of men and even women cluster at different points along the way, debating or listening to speakers expounding on philosophical, political, and religious matters.

Dion stops and listens to different speakers, but then comes to a square where a large crowd has gathered. They appear to be mostly Jews.

"Who is it that draws such a large group?" Dion asks a bystander.

"Philo Alexandreus," the man replies.

Philo of Alexandria, Dion recalls, is a Jewish philosopher who bridges the periodic religious strife between the Greek and Jewish quarters of the city. Philo walks between the two worlds, synthesizing Greek philosophy with Jewish religious beliefs. It's not long before Dion is spellbound.

"The Hebrew Bible has been misunderstood, particularly by the Hebrews," Philo said. "Any sacred text worth studying cannot be taken literally because, no matter how much literal truth lies therein, the allegorical truth is far greater. Allegorical interpretation indicates that God has created ideal forms and archetypes that are the patterns for a perfect creation. Yet, we do not have perfection."

His words set some murmuring off in the crowd, but Philo continued undeterred. "So it is that you see God in the image of an earthly king dispensing punishment or rewards, or commanding sacrifices from His people. I ask this question—how is it that a merciful, all-powerful God created such suffering and imperfection as we see all around us?"

The people look around, as though the person next to them might have the answer they do not possess.

"If," Philo said, "we hold that God is good, kind, and merciful, and if God has created a world of perfect forms from which the material world arises, then something stands between us and God. Pierce that veil and you will see the true reality of existence."

Roman soldiers then walk into the square and the crowd begins to disperse, but not before Philo's words rub together like flint in Dion's mind, sparking the fire of curiosity.

A Path Is Set

A few weeks after Philo's oratory, Dion attends an informal gathering of Greek-speaking Jews belonging to a circle that follows Tryon, a local Greek philosopher. At one point, Dion speaks up.

"I heard Philo speak recently. Much that he said makes sense to me. Where does one acquire such knowledge, and with so many different teachings abounding, how can we know which knowledge is true?"

"You will never get truth from anyone else, only direction," Tryon says.

"Then how do we learn the truth?" Dion asks.

"By direct experience only," Tryon replies. "Anything else is taken on faith, taken on someone else's word, and you have no way to determine its validity."

"I wish to have such experiences," Dion says. "How do I come by it?"

"By hard work, long study, sacrifice, and contemplation," Tryon says. "Are you ready for that?"

"If I can learn to perceive like a Plato, a Pythagoras, or a Philo, I will do anything," Dion declares.

"Is that so?" Tryon says. "Come here next week. We'll see if you are suitable for a certain place I have in mind where they can help you, but don't get your hopes up. This is a serious path we speak of, not for everyone, but come and we shall see."

Dion waits in anticipation the entire week, elated but also nervous, for he senses he might be entering a new world, and this world will not come easy.

Luxor

Dion has just completed his first interview at the Tentyra temple, thirty miles from Luxor, in Upper Egypt. He came for the interview three days in a row and waited all day without being seen. He tried not to show anger or frustration, sensing this might be some test of his will or desire. On the fourth day, an advanced adept named Thutmes meets with him.

Figure 13. Temple of Luxor

Thutmes asks him a series of questions about his background, his intentions, and his expectations. He shows Dion a series of drawings, some seeming nonsensical, and asks him to interpret them in his own words.

Thutmes then presents him with some mathematical problems to solve and, owing to his Greek education, Dion does so. Dion undergoes other forms of questioning, and is then told to wait until he is contacted.

A few days later, to his great delight and excitement, Dion is accepted into the school. He feels like a young voyager on his first ship, looking out onto the wide ocean and sensing a great unknown adventure lying ahead.

Seeing the Light

During Dion's seven years in the temple, he attends classes in mathematics, geometry, philosophy, and harmonics to sharpen his intellect. After completing these levels, he finally meets a Phoster—this title means "illuminated one."

The Phoster tells him to forget everything he has just learned. He now has to pull himself out of his mind and into his heart. He has to stop thinking and start feeling.

"Do not worry," the Phoster says. "All your book learning is stored here," and he taps his temple. "It won't vanish. But the world came into being by the two polar forces of the Great Mother and the Heavenly Father. You can only bring these into balance by first allowing one, then the other full sway within your being, so that you may know them in the purity of their inner nature. Only then can you merge them."

Dion spends many hours of contemplation in the same windowless room for several years. Sometimes he is in complete darkness, other times with a solitary candle with which to meditate.

Sometimes he is in complete stillness, other times the Phoster creates tonal vibrations from different instruments that resonate throughout the room. His powers of concentration and visualization sharpen. He learns incantations and is given certain talismans and pictures of holy sages upon which to focus. He has visions, which he records and discusses with his Phoster. At first, he is trained to use these visions to solve practical problems to reinforce that the information he is receiving is "real."

He begins to know things he could not learn from his five senses. He senses when a person is ill with no outward sign of sickness. He sees events and knows conversations before they come to pass. These experiences ground him in the knowledge that humanity and the world are encased in a web of subtle energies that govern creation.

The time comes when he has his first guided meditation with his Phoster. Dion loses all sense of space and time. At some point, his being is engulfed in a pearly, opalescent light, and the light speaks to him, speaks to his mind.

But then Dion perceives that it doesn't really speak to him; it speaks through him because he is part of it. He is not separate from it. What Dion thinks to be his identity is merely a limited aspect of what now informs him and bathes him in knowledge. And from this day forth, he realizes he now walks the world with opened eyes . . .

14

Achieving Gnosis

The Adept Initiated

Judea

It is now 31 CE. Dion has lived to the ripe age of forty-two, thanks in part to the dietary regimen he learned at the Tentyra temple. He has traveled to the Roman Judea province, having heard about a young rabbi named Yeshua of Nazareth.

Some said this Yeshua had also been an adept at one of the Egyptian temples, even undergoing the resurrection experience, the highest and most dangerous level an initiate could attain. Many miraculous things had Dion experienced, but he had never attained the resurrection level.

News of Yeshua reached Alexandria through the Jewish community there. Some, held to be wise men, whispered that Yeshua was from the line of redeemers, an expected one bringing sacred knowledge to an oppressed humanity in danger of falling back into barbaric darkness.

Figure 14. Computer generated model of Ancient Jerusalem

Upon arriving in Judea, Dion attends one of Yeshua's public sermons. This is certainly a wise man, he thinks, but no profound revelations does he hear. Dion stays until the crowd disperses. About twenty people surrounded Yeshua, equally divided between men and women. Dion notices a lovely woman, possessed of lustrous reddish-gold hair.

"My lady," he says, "I have listened to the words of Rabbi Yeshua. Yeshua is surely learned, but I sense there is much he holds back, as if his water flows through a sieve but his gold is left in the basket. He gave water to the masses today, but I would experience the golden wisdom he leaves hidden. You are a follower of his, no? Are my impressions not true? For, if they are, I would hear his words unfiltered."

The woman's gaze pierces Dion like an arrow through his heart. "You, sir, show the discernment of the wise. Come back in a fortnight to the grove beneath yonder hill," she says, pointing to the east. "I shall see you are welcome."

"My thanks to you, dear lady. How may I call you?"

"My name is Mary of Magdala."

The Living Light

The night is calm and the grove fragrant, two weeks later, as Dion sits among a group of thirty men and women. He already knows something is special about this rabbi and this group, because so many women are present.

Teaching women in public in mixed company is a grave offense under Jewish law, and Dion understands why they chose this place and time to meet. Mary of Magdala stands at Yeshua's side. Sometimes, during his discourse, she rephrases a question to him in a way for the answer to be better understood by the assembled. Other times, she elaborates on a point he has made, always to make for the greater clarity of information.

"This earth is ruled by those who neither hear, nor see, nor comprehend and yet by those is held all the power of this world, yea even the power of life and death," Yeshua says.

Figure 15. Jesus and the Disciples

"Do you speak of the Romans, Rabbi?" a follower asks.

Yeshua extends his hands, palms outward, to his audience. "You think the Romans the oppressors and you the oppressed?" he asks as he hammers his fist into his open palm. "I say both you and the Romans are the oppressors and the oppressed." His voice is firm, and it cuts through the stillness of the night, but the torchlight reveals a gentle smile on his face.

"Every nation has ruled and been ruled because people believe in a God who behaves like an Earthly monarch, looking for obedience and dominion, and dispensing justice for good and evil. You have created the Romans, the Greeks, the Assyrians, yea even the Hebrews as they took the land of Canaan, for you shape God with the frailties of man when you should shape man with the glory of God."

Dion grasps Yeshua's meaning, but many look bewildered.

Then Mary of Magdala speaks. Her voice is steady, calm, and reassuring. "To understand the Master's words, you must understand your own origins and the nature of God." She looks at Yeshua, and he nods.

"You all have the ability to conceive a thing and manifest it," Yeshua says. "You may think of a chair, and then you build it according

to your mind's vision. In this ability, you reflect the nature of God, for the vision of the One Great Mind conceives and sustains all things."

"God is the First Thought," Mary says, "and the First Thought has dispersed Itself. Your mind, your ability to think and conceive things, derives from God's very essence. Just as you are formed by the thoughts of God, so you think and form the things of the material world."

"But we did not form the sky or the oceans," someone says.

"Your existence is like the steps on a stairway," Yeshua says. "You exist as spirit, soul, and body. You create at each level of existence. The world you see was fashioned at higher levels. At those levels of spirit and soul did you participate, each adding some small measure of yourselves in the creation of all you see."

A flood of questions ensues, bursting forth like an overflowing dam: Why are we not conscious of this power? Is this God you speak of the God of the Hebrews? How can we be of God's essence when we are imperfect? If all things come from God, why does so much evil exist? You have said we labor under the yoke of the Archons—who are these powers?

Yeshua smiles. "We sit here beneath the trees. Imagine instead, you are seeing the sun from yonder clearing." He points to the open plain behind them. "Imagine how you would see the light shining, bright and unobstructed. But you are removed, you are here in this grove. You cannot see the sunlight for the trees. You receive light, but it is filtered with shadows."

"So," one follower says, "you are saying a veil exists between us and God—this is symbolized by the trees that filter the light. But who creates this layer of shadow? Who are the beings that oppress us? Are these the Archons of which we hear spoken? Do they create the evil in this world?"

"Can any among you answer these questions from that which I have previously given you?" Yeshua asks.

Dion raises his hand. "May I demonstrate with some stones and a stick, so that things may be better grasped?" he asks.

Mary gives him permission to, and the people assemble around *him. He gathers several stones and a stick with which to draw in the dirt. He draws four circles and arranges the stones.*

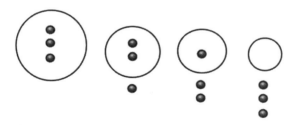

Figure 16. Consciousness Contracting to Project Creation

"The circles," he says, "represent God, the Totality of all that is. The three stones within the circle on the left represent the potential for individual consciousness. God now contracts Itself, as we see in the second circle to the right, so that one stone is now outside the circle. This is how God limits the presence of Its all-encompassing consciousness to allow individual identity to come into being. As God continues to contract Itself, another stone appears in the space outside the circle, and so on until a chain of stones is projected outside of the Whole."

Yeshua has a warm smile, and he nods to Dion, urging him to continue.

"The individual stones outside the totality are made of the substance of the Whole from which they came," Dion says. "But now they are in a different dimension, apart from the place of their origin, and the farther away each stone, or soul consciousness, is from its Source, the more ignorant it is of the Source. Thus we are sparks of the divine lost in the forgetfulness of ignorance. The shadow from the Master's parable of the forest comes from us."

Mary speaks, also giving Dion a smile of approval. "Our brother speaks well. Shaded from the One Light, our diminished awareness creates a world more of our making and less of God's," she says. "The world is imperfect because it comes from the many, not the One. We became our own oppressors. But it is illusory—it is a dream from which we can awaken."

The people begin to talk amongst themselves, in wonderment at the information just given, until Yeshua again speaks.

"Evil is the extreme ignorance of those who have forgotten their unity with the Source. In selfishness, they do injury with no understanding that their actions ripple throughout Creation. But one day, like children at play in the fields of the Lord, we will set aside childish things and mature back into the full light, bringing the experience of our creation with us."

Yeshua continues, "Indeed, you are the fingers of God touching the face of this world. Your purpose is to spiritualize the material, and bring the experience of the material back to spirit. In this way, God has chosen to experience Itself in all dimensions of expression."

After the teaching, Dion speaks with Yeshua. "Master, before I came into your presence, I would never have been able to articulate things I have experienced to others, as I did this night. I know it is you that has drawn this forth from me."

And Dion stayed years in Judea, deepening his knowledge at the foot of the Master until one day he departed to spread the good news to his homeland. For, if Yeshua was a flame, those like Dion were embers blowing in the wind to ignite the lineage of light around the world.

Darkness Rising

Dion, blessed with an unusually long life, is now an improbable eighty years of age, but he is fraught with great sorrow. The mystic Way of Jesus, now called Christianity, spread like wildfire, but it is no longer the way of Jesus. After the Nazarene's death, the female leaders of the Way, such as Mary of Magdala, were killed or pushed aside.

The mystical body of seekers has degenerated into a hierarchy of men with limited understanding and self-serving agendas that have distorted Yeshua's revelations with twisted dogmas he never taught. They bend the truth to become acceptable to the state and gain Roman favor. Great light attracts great darkness, and the forces of opposition will ever blind and delude people with false ideas claiming to hold truth. This age, Dion realizes, is not ready to grasp the great mysteries.

Like Camelot, the mystical knowledge of Yeshua and the female disciples burned for one brief, shining moment during the darkness of Roman domination. The mystery schools would vanish from the Earth, the sacred knowledge would be suppressed for ages. The minds of men would turn from the heavens to fixate on dominating the Earth.

But if Dion were alive today, he could take some measure of consolation. The sacred knowledge kept reincarnating in the most unusual of forms from the tarot to troubadours, freemasons to recovered gospels, and now to findings in the most advanced scientific laboratories.

The great eternal story of our existence keeps unfolding before our eyes.

15

What We Lost

The Feminine Path of the Gnostics

T he dominance of Mary Magdalene as a disciple and Dion's dismay at the changes in Jesus's teachings underscore a tragic episode in Western spiritual history. It was tantamount to the suppression not only of historical female spiritual leaders but of the body of mystical wisdom produced by adherence to the feminine principle as the intuitive gateway to enlightenment.

It's an undisputed fact that Gnosticism and Jewish Kabbalistic mysticism are closely related, both being branches of the ancient mystery school teachings. On their face alone, their Creation myths are very similar. These common, sacred traditions predated Jesus and immersed the culture of his time.

We have evidence of direct links between Jewish sects and Gnosticism. Church father Epiphanius, writing in the fourth century CE, speaks of the above-mentioned Nazoreans (Nazarenes, Nasoreans) and Ossaeans who existed before Jesus. They were Jews who believed in a different Mosaic revelation, and are distantly survived today by practicing Gnostic Mandeans, Semitic people living in Iraq and Iran.

In chapter 11, I shared writings from church fathers and the Bible that clearly indicate that Jesus, like Moses and so many other spiritual figures, passed on a secret oral teaching to his disciples. Thus, two streams of Christianity initially existed, one stemming from the secret teachings and one from the parables of the public teachings. The public teachings formed the basis for the mainstream Christian churches of the present day.

But what were these secret teachings? Given all the information presented so far, and the direct evidence in the suppressed Gnostic gospels, Jesus was surely imparting a mystical Gnostic or Gnostic-Kabbalistic wisdom.

Even absent the content of his old oral teachings, any average scholar would be able to make a strong circumstantial case for the mystic Jesus. He lived in the shadow of Mount Carmel, long home to an ancient Essenic Mystery School. He spent years in Egypt, the center of the ancient mystery tradition. Early orthodox church fathers admitted that he taught a body of secret mysteries. And the Gospel of Thomas, one of the earliest gospels yet found, is full of Jesus's teachings of Gnostic wisdom.

FEMALE LEADERS OF THE EARLY WAY

The terms "Christians" and "Christianity" did not exist during Jesus's ministry. His particular mystical teachings were called the Way by his followers, and many early leaders of that Way were women. Jesus was the radical feminist of his time. You have to understand just how suppressed women were in the Jewish culture of his age. It was unseemly to be talking to women in public, particularly unmarried women, let alone have them act as leaders within your group.

Women of that time could not inherit land or choose divorce. They were not educated in the Torah or allowed to pray in public. They were disqualified from being witnesses in legal disputes. When menstruating, they were considered unclean, so when Jesus allowed the hemorrhaging woman to touch him in Mark 5:25, it demonstrated his radically different attitude toward women.

The New Testament gospels make it clear that women were among Jesus's earliest followers. They mention names like Mary Magdalene, Joanna, and Susanna as women of means who accompanied Jesus during his ministry and materially supported him and his poor disciples (Luke 8:1–3). Among these, Mary Magdalene is of particular prominence.

MARY MAGDALENE

Mary Magdalene is the mysterious figure who reads, in the Bible, like an artist's masterpiece painted over by a new painting. Her outline is still visible around the edges, like she refuses to be erased, given her critical role in the original Way. She is constantly at Jesus's side, even at the crucifixion after the male disciples fled. She is the first to see the risen Jesus, and delivers his message to the disbelieving disciples so that even the church had to confer on her the title Apostle to the Apostles.

Figure 17. Jesus and Magdalene

The Gnostic gospels are far more revealing of the Magdalene story than the Bible, which was edited by orthodox opponents of female mysticism. Mary is unequivocally considered Jesus's most important and brightest disciple. She clarifies Jesus's teachings to others, and conversely frames questions for others to ask of him.

St. Peter, who is tagged as the primary disciple in the Bible, is a jealous dunce in the Gnostic gospels. He has a hard time grasping the mystical teachings, frequently complains about Magdalene's dominance, and is expressly jealous of her, asking Jesus why he loves her above the other disciples. Any doubt about Mary's standing (also called Mariam or Maria, from the Hebrew) is put to rest by the following passages from the Gnostic gospels of the Nag Hammadi Library:

- From the Gospel of Thomas: "Simon Peter said to him, 'Let Mary leave us, for women are not worthy of life.' Jesus said, 'I myself shall lead her . . .'"[1]
- In the Dialogue of the Savior, she is described as a woman "who had understood completely." Jesus says Mary makes clear "the abundance of the revealer."[2]
- The Gospel of Philip says ". . . the companion of the [Savior is] Mary Magdalene. [But Christ loved] her more than [all] the disciples, and used to kiss her [often] on her [mouth]. The rest of [the disciples were offended] . . . They said to him, 'Why do you love her more than all of us?' The Savior answered and said to them, 'Why do I not love you as [I love] her?'"[3]
- In *Pistis Sophia* Jesus said, "Mariam, thou blessed one, whom I will complete in all the mysteries of the height, speak openly, thou art she whose heart is more directed to the Kingdom of Heaven than all thy brothers."[4]
- In *Pistis Sophia* Jesus said, "Excellent, Maria. Thou art blessed beyond all women upon Earth, because thou shalt be the pleroma of all Pleromas and the completion of all completions."[5]

Besides the strong presence of women, Jesus's Gnostic teaching was oriented toward the feminine principle. As I've pointed out, it's the female qualities of intuition, imagination, and receptivity that open the gates of consciousness, especially in the face of rigid, male, patriarchal thinking and customs. The difficulty of Peter and some of the male disciples to break out of the box of Jewish Law and grasp Jesus's teaching indicates the struggle of the male disciples to transition their thinking through this intuitive mindset.

SPIRITUAL HOLOCAUST

The spiritual message Jesus and Magdalene taught was powerful, but it bucked the tide of the completely patriarchal world of the Jews and the dominant Greco-Roman culture of that age. Its spread was also hindered by the constraint of the mystic tradition, which held that complex spiritual teachings were not for mass consumption. This restriction was not from elitist tendencies but from fear that the masses would not understand and the wisdom would be corrupted.

If the comprehension of the disciples in the Gnostic gospels is any indication, that fear was well founded. In uninitiated hands, the teachings would be bastardized and twisted, which is rather what happened anyway. Simultaneously, an outer doctrine was beginning to develop stemming from Jesus's public teachings and parables like the Prodigal Son. The warping of the teachings came about perhaps partly from misunderstandings transmitted by the male disciples themselves, but certainly through the mission of Paul of Tarsus.

St. Paul never met Jesus, and what he taught differed from what Jesus taught. Paul's preaching was an odd mix of Gnosticism, misunderstood public teachings, and seemingly his own invention. It was this twisting of the sacred wisdom that the mystical schools had feared. Despite Paul's ascendancy, Gnostic Christian communities thrived for some time, though with increasing estrangement from the Christian branch known as Orthodox, coalescing around the Pauline teachings.

As time passed and the sect that became known as Christians grew, they experienced terrible persecutions by the Roman Empire. The only way to stop the oppression was to be accepted by Rome or, better yet, adopted as the Empire's official religion. But they had a big problem—women. From their origins, the movement had a strong female component of leadership, even more so in the earliest Gnostic-oriented communities.

Christianity wasn't the only religion competing for primacy in the Empire, but it was the only one with so many female leaders. This was an embarrassing liability in a patriarchal world. Thus, the movement to push female leaders and their teachings out of the mainstream got underway.

Part of this effort may have been to conflate Mary Magdalene with the disciple John, to cover her legacy without completely expunging the history of so important a Christian figure. In other words, don't do away with her altogether, just make her into a man. Many of you are probably familiar with the very female-looking figure next to Jesus in Da Vinci's

Figure 18. The Last Supper

The Last Supper. Many have identified that figure as Mary Magdalene, as was popularized by Dan Brown's *The Da Vinci Code,* but she was historically passed off to the world as St. John. John also happened to be the sole male disciple with Magdalene at the crucifixion. Coincidence?

We have the exceedingly odd passage in Mark 14:51, where the Romans have come to seize Jesus and the disciples have fled. The last person with Jesus was said to be a young man:

A young man, wearing nothing but a linen garment, was following Jesus. When they seized him, he fled naked, leaving his garment behind.

This episode makes far more sense if the young man was really Mary Magdalene, the constant companion of Jesus, always the first and last around him. Finally, Ramon Jusino, M.A., wrote a very compelling essay, which is still available online, stating that the Gospel of John was originally the Gospel of Mary Magdalene. This would certainly account for the differences between John and the three synoptic gospels of Matthew, Mark, and Luke.

Magdalene's stature would have been second only to Jesus. The seeds of misogyny, however, were evident early on, as we know that even male disciples like Peter despised her. The move to make her a man fits the idea of taking an eraser to Magdalene while leaving traces of her great influence, like painting over a masterpiece with another work. Recall how Leon Trotsky was second only to Lenin in the Bolshevik revolution, and headed the victorious Red Army, while Stalin was a flunky. Yet Stalin eventually twisted, then eradicated, Trotsky's legacy.

Great lies by historical victors are easily adopted. The male hierarchy of the Orthodox were victorious, but at the cost of Rome altering the Way in its own image. Erasing the mystical female legacy and replacing it with a Roman church hierarchy more interested in political control than spiritual growth wasn't too difficult. For visual evidence of this eradication, see the fresco in figure 19, found in a cave near the former Greco-Roman city of Ephesus, in modern-day Turkey.

Figure 19. Cave Mural, Ephesus

The two figures are St. Paul and a famous female saint, Thecla. Notice that Thecla is higher than Paul. This was significant as, in those days it meant higher authority. Also, note the gesture they're making with their hands. It's called the sign of the bishop, and is still used in Catholic and Orthodox churches. It indicates a person of authority, someone who is learned and revered.

But see what happened in ensuing years—Paul's image remains untouched while Thecla's eyes and fingers are chipped out, clearly a directed effort to disrespect and eradicate her authority. This image is emblematic of what happened to female Christian spiritual figures in general. You don't have to eliminate them, just de-feminize and fade them out, like Magdalene.

WHAT WE LOST

The mystical teachings were burned or otherwise suppressed. In their place, we got a pile of guilt-producing dogmas that Jesus wouldn't recognize, like original sin and the separation of man from God. The Holy Spirit, once decidedly identified as a feminine force in early Judeo-Christianity, was neutered to become an "It."

Scholar Karen L. King, in her book *Women In Ancient Christianity: The New Discoveries,* shows what Christianity looked like from the feminine perspective:

- Jesus was understood primarily as a teacher and mediator of wisdom, rather than as ruler and judge.
- Theological reflection centered on the experience of the person of the risen Christ more than that of the crucified savior. Interestingly enough, this is true even in the case of the martyr Perpetua. One might expect her to identify with the suffering Christ, but it is the risen Christ she encounters in her vision.
- Direct access to God is possible for all, through receiving the Spirit.
- In the Christian community, the unity, power, and perfection of the Spirit are present now, not just in some future time.
- Those who are more spiritually advanced give what they have freely to all without claim to a fixed, hierarchical ordering of power.
- An ethic of freedom and spiritual development is emphasized over an ethic of order and control.
- A woman's identity and spirituality could be developed apart from her roles as wife and mother (or slave), whether she actually withdrew from those roles or not. Gender is itself contested as a "natural" category in the face of the power of God's Spirit at work

in the community and the world. This meant that, potentially, women (and men) could exercise leadership based on spiritual achievement apart from gender status and without conformity to established social gender roles.

- Overcoming social injustice and human suffering are seen as integral to spiritual life.

A PAGAN SIDEBAR

Before we begin our examination of ancient Christianity, we should clarify an important point here concerning the emotionally charged word *pagan*. This word was created, and acquired prejudicial meanings, as a result of the triumphant narrow orthodox beliefs of the three Abrahamic religions: Judaism, Christianity, and Islam.

Look at the dictionary definitions of the word:

1. One of a people or community observing a polytheistic religion, as the ancient Romans and Greeks.
2. A person who is not a Christian, Jew, or Muslim. Synonyms: heathen, gentile, idolator, nonbeliever.
3. An irreligious or hedonistic person.
4. A person deemed savage or uncivilized and morally deficient.

Heathens, idolators, savages, and morally deficient—given the facts and histories of organized religions, that's the pot calling the kettle black. The variants of the universal ancient mystery school tradition, as we've seen in this series, represented the most insightful, advanced, and even scientific approach to answering the deepest questions of life and Creation in human history. The major religions are a pale externalization of these esoteric traditions, designed for mass consumption and thereby vulnerable to manipulation by powerful people with personal agendas over the ages.

When the Greco-Romans waged war, at least they didn't justify it in the name of Zeus, unlike Christians and Muslims who killed in the name of Jesus or Mohammed. That's one of the main things that distinguishes religion from spirituality—the "my God is better than your God," "I'm right and you're wrong" mentality. So much for moral deficiency.

HYPATIA

Hypatia was a brilliant Alexandrian Greek Neoplatonist philosopher, astronomer, and mathematician living in Egypt at the time when the Orthodox Christians were gaining ascendency in the Roman Empire. Neoplatonism was very close to Gnosticism, and Hypatia was affiliated with the Serapeum, the remaining repository of universal "pagan" wisdom surviving the tragic burning of the great Library of Alexandria.

Figure 20. Hypatia

Theophilus, the Christian bishop of Alexandria, had the Serapeum burned. This act set Western civilization back in ways that are unimaginable to this day. The countless Greco-Roman and Egyptian classics that went up in flames would have altered the course of history if preserved. Hypatia was symbolic of the "pagan" scientific and mystical wisdom aided by the spirit of free inquiry from which Jesus's secret teachings had stemmed. Not long after the Serapeum incident, Hypatia was brutally murdered by Orthodox Christians looking to exterminate all things pagan.

Her murder marked the end of the classical pagan era of science, wisdom, and open inquiry that so characterized the feminine orientation to solving life's mysteries. The rule of dogmatic, male hierarchies would now seize control of the Western world with a repressive hand in the name of religion.

This deviation from our story is important so that the reader does not reactively dismiss the power and validity of the ancient wisdom tradition because of prejudicial meaning attached to the word *pagan*. Religions like Christianity were not just plopped into history in a vacuum. They owe their existence in large measure to distorted, mass-produced understandings of this much older and wiser spiritual tradition called paganism.

The strange fact of modern existence is that we seem to progress in science but regress in spiritual matters. So, if anything, our ancient ancestors have at least one facet of life in which they might claim superiority over us. Please keep this in mind whenever you see the word *pagan*.

THE GNOSTIC HOLOCAUST

It must be obvious by now that the Gnostic and mystical traditions in general didn't escape the heavy hands of the same ilk that murdered Hypatia and destroyed the Serapeum.

The process of the diverging Gnostic and Orthodox Christians accelerated under the Roman persecutions. The hierarchical, institutionally minded Orthodox Christians had to come up with a real survival strategy. They switched gears from merely spreading their version of the Christian story to becoming the official state religion of the Roman Empire. The best way to beat them was to join them, and they ultimately succeeded. The agenda was now theirs to set.

By 325 CE, the politically and militarily astute emperor Constantine saw his empire fragmenting politically and administratively. But Constantine and his mother Helen were shrewd. They recognized the power of religion to unify and snap people into line. After all, God-driven ideologies are incredibly powerful. Even today, we see this in Islamic terrorism, where young people rush to blow up themselves and others to gain God's grace. There's no fanatic like a religious fanatic.

Constantine assembled an all-male cast of bishops at a place called Nicea, and commanded them to come up with a creed or dogma prescribing what it meant to be a Christian—and it had to meet his approval. Up to this point, Christianity existed in far-flung communities, from India to the British Isles. Each community had its own variant of the Christian story, focusing on what was important to them.

After 325, this was no longer the case. Only one way of believing was allowed, under threat of excommunication—or worse. The corporate, mass-market version of Christianity had triumphed, further altered by the dictates of the Roman Empire, to become Orthodox Catholic Christianity.

COME, THE HERETICS

The trend to demote women and alter or expunge their works from history was already underway before 325. But now an official hierarchy existed, sanctioned by Rome to eliminate as heretic any who would not toe the new party's line.

This included relatively Orthodox Christians, such as Arians, with slight nuances of difference over certain views of Christ's nature. And it certainly included those hard-to-comprehend mystics and their female leaders. The newly consecrated public church now turned on its own original, mystical core with a vengeance.

After Constantine's council, the days of the Gnostic Christians were numbered. At first, they were thrown out of the churches; but as the new Orthodox Catholic church became more organized, authoritarian, and dogmatic, things got worse. The Gnostics were persecuted and killed, and their writings burned. The burying of the Gnostic gospels in the Egyptian caves, near the still-standing Greek monastery of St. Pachomius likely occurred when a historically known bishop's decree from Alexandria ordered the destruction of the newly heretical mystical texts. Apparently, some enlightened monks disagreed, and hid the papyrus texts in earthen jars to lay dormant for two thousand years.

Before the discovery of those buried texts, the destruction of the mysteries was so complete, modern scholars had to rely on the writings of hostile church fathers, like Irenaeus and Hippolytus, to glean something of Gnostic beliefs.

The fate of the Gnostics, the mystical church, and the female leaders was the same because *they were the same.* The last organized church that could be called Gnostic cropped up in southern France, in the eleventh through the thirteenth centuries. This was the church of the Cathars. Their name came from the Greek word for purity. They actually called themselves the "Bon Hommes," meaning "good people" or "good Christians."

The Gnostic influence supposedly came to France from Bulgaria via Armenia, though persistent legends of Mary Magdalene living in the heartland of the Gnostic Cathars are quite visible to this day. The Cathars believed they taught the original ways of Christianity from before the domination of the Roman church. They practiced healing and vegetarianism, and roamed the countryside assisting the needy. Their priesthood was almost entirely female, and the female priests were called *parfaits,* or "perfects."

They held the Church of Rome to be in error, and called themselves the Church of Amor (love), which is "Roma" spelled backward. All of Provence and southern France were Gnostic—nobles and peasants alike—and southern France was wealthy.

The Cathars were the most prominent sect in a confluence of Gnostic revivals at that time, in a region that included the Islamic Sufis in Moorish Spain, the troubadours of southern France, and some smaller movements in northern Italy. The pope resented Cathar independence and their criticism of the church, while the king of France and the northern French nobles coveted the Cathars' wealth.

This greed and animosity was the basis for the formation of the only Crusade launched by Christians against other Christians: the Albigensian Crusade. Throughout the twelfth and thirteenth centuries, the Gnostic Cathars were destroyed in a most brutal fashion.

Figure 21. Persecution of the Cathars

They were hunted, tortured, and burned alive; their lands were stolen and their works destroyed. The tone was set very early in the Crusades, when the Cathar stronghold of Beziers was under siege by the Crusaders, in 1209. One soldier in the papal army asked Pope Innocent's legate, Arnaud Amaury, how to tell the innocent from the guilty, as the city was also home to many Catholics. The legate's chilling reply: "Kill them all, God will know his own."[6] The popular rendition of this phrase is: "Kill them all and let God sort them out." Thirty-five years later, the destruction of the Cathars was almost complete, with the capture of the fortress of Montsegur and the burning of the Cathars captured there who refused to renounce their faith. The destruction of the Cathars ended the last expression of large-scale organized Gnosticism. From that point on, their knowledge would go underground, finding outcroppings in the tarot, Rosicrucianism, Freemasonry, Theosophy, and even Mormonism. The modern New Age movement was largely fueled by understandings (and misunderstandings) stemming from Gnostic wisdom.

But people will always be able to perceive wisdom and truth. It dwells in our hearts, and no manner of persecution can fully eradicate that. The Gnostics have a story to tell, and once that story is understood, it will rock the Western world and free our spirits from the stagnation of our religious and anti-religious boxes.

16

Journey of the Mind Voyager

Psychological Discoveries of Carl Jung

Ironically, a famous twentieth-century psychologist would play a major role in reintroducing Gnostic-feminine mysteries into mainstream thought.

Ancient Gnostic wisdom took the form of myths. Since myths are symbolic, people will see them as whimsical fantasy unless their inner messages are deciphered in more concrete terms.

We previously explored the spiritual, psychic, and material dimensions. We also analyzed how they manifest as spirit, mind (soul), and body.

DIMENSION	MANIFESTATION	DISCIPLINE
Spiritual	Spirit	Mysticism
Psychic	Mind	Psychology
Material	Body	Science

Figure 22. Dimensional Manifestations

The methodologies we use to describe these concepts are listed in the right-hand column of figure 22. Mysticism deals with the spiritual dimension and spirit. Psychology deals with psychic (mind) functions. And science (physics) describes the physical world. Next, we'll deal with the psychic dimension.

The psychic dimension is the intermediate state of being, the bridge between the spiritual and material planes. Carl Jung, the eminent founder of modern depth psychology, recognized this.

> Between the unknown essences of spirit and matter stands the reality of the psychic—psychic reality, the only reality we can experience immediately.[1]

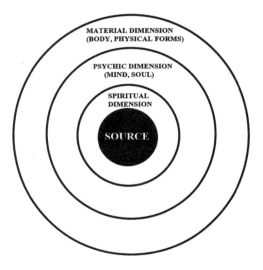

Figure 23. Dimensions Radiating from Source

The diagram in figure 23 depicts the outward radiation of the Source to the spiritual, psychic, and material dimensions respectively, with the psychic bridging the spiritual and material.

CARL JUNG

In this chapter, we'll learn the psychological tools produced by the pioneering work of the great Swiss psychologist Carl Jung. We'll then examine the Gnostic myths through the lens of the modern analytic (depth) psychology Jung founded.

Jung brought us concepts and terms that have become staples in psychology and popular usage, such as synchronicities, archetypal phe-

Figure 24. Carl Gustav Jung

nomena, the shadow, the collective unconscious, psychological complexes, extraversion, and introversion.

The interesting thing about Jung is that he was kind of a "spontaneous Gnostic." Some people have flashes of clarity, transcending our virtual matrix to reveal reality-changing sources of higher information. Philip Dick, author of the adapted movie *Blade Runner,* is an example of one who strongly felt he had such insights.

Jung's discoveries did not come strictly by observation but by introspection, that is, exploration of the workings of his own mind. Jung underwent certain personal experiences, then later correlated them with observable phenomena in other people and cultures. He was astonished to learn that the Gnostics had preempted many of his discoveries three thousand years earlier. Upon reading the Gnostic texts, he remarked, "My whole life I have worked to know the soul and these people [Valentinian Gnostics] already knew it."[2]

Gnostic scholar G. Filoramo wrote: "Jung's reflections had long been immersed in the thought of the ancient Gnostics to such an extent that he considered them the virtual discoverers of 'depth psychology.'"[3]

JUNGIAN PSYCHOLOGICAL CONCEPTS

Before we look at the Gnostic Sophia myth through the eyes of Jung's depth psychology, we'll summarize some of the key concepts Jung introduced to the field of psychology.

Unity/Polarity

As in the Gnostic Creation story, Jung recognized that the ground state of being (Source) is a unity of polar aspects:

> So the union of the two is a kind of self-fertilization . . . He divided his self [Atman, God] in two, and thence arose husband and wife.[4]

Both Jung and the Gnostics believed that these male-female polarities became pronounced at lower levels of existence. Jung called the female aspect *anima* and the male aspect *animus*. They are hardwired into our cognitive functions, originating in separate brain hemispheres. They manifest as left brain and right brain functions, as shown in figure 25.

Jesus was attributed in the Gnostic texts as saying these two orientations within us must become one again to realize the Kingdom of Heaven, or divine consciousness:

> When you make the two into one . . . that is, to make the male and the female into a single one . . . then you will enter [the kingdom].[5]

LEFT BRAIN FUNCTIONS
uses logic
analytical
detail-oriented
facts rule
words and language
math & science
knowing
reality-based
forms strategies
practical
safe

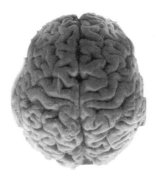

RIGHT BRAIN FUNCTIONS
uses feeling
intuitive
"big picture" oriented
imagination rules
symbols and images
philosophy & religion
believing
fantasy-based
presents possibilities
impetuous
risk-taking

Figure 25. The hemispheres of the brain

Jung said the same thing:

> Wholeness is not achieved by cutting off a portion of one's being, but by integration of the contraries.[6]

Jung further criticizes the exclusively logical male approach to life to the detriment of our feminine intuitive faculties:

> We should not pretend to understand the world only by the intellect; we apprehend it just as much by feeling.[7]

Individuation/Attaining the Self

The Self is the source of the whole range of psychic phenomena, as experienced by humans. Both Jung and the Gnostics strove for the same end goal—to awaken the spark of God lying dormant within us. The aim of both Jung and the Gnostics was to throw off the illusions and the *counterfeit spirit,* the layers of ego, false personas, ignorance, and deceptions that keep us from the destiny of regaining our higher (divine) state.

Jesus says throughout the various Gnostic gospels that people are asleep, that they have neither the eyes to see nor the ears to hear the wisdom of their own higher selves. Jung says the same thing when he writes that we have to integrate our chaotic unconscious minds to awaken our true selves. Witness these juxtaposed quotes from the Gnostics and Jung, respectively:

> If you bring forth what is within you, what you bring forth will save you. If you do not bring forth what is within you, what you do not bring forth will destroy you.[8]

> When an inner situation is not made conscious, it appears outside as fate.[9]

What these quotes tell us is that our higher consciousness lies buried beneath our conscious and unconscious minds. If we access it, we can write our own life's script. If not, the script writes us and we are driven to our detriment by less conscious impulses that appear to us outwardly as fate or misfortune.

Collective/Personal Unconsciousness

Just as we have a personal unconscious mind, where individual memories and repressed experiences are stored below our awareness, all humans share a collective unconsciousness. For Jung, this was kind of a racial memory, where humanity stores its collective past experiences as spiritual, mythic, or religious symbols.

Archetypes

Archetypes are images or ideals, arising from the collective unconscious, that translate into patterns of behavior. For example, the mother figure is the archetype of a nurturing person. But such archetypes can be both positive and negative. The mother who suffocates, mistaking this for love, is a negative distortion. Jung evolved from looking at archetypes as a biologically based phenomenon to understanding them in the Gnostic sense of having a real existence as centers of consciousness (Aeons) in an objective, independent psychic realm.

Jung suggested that the archetypes govern human behavior, and even the behavior of inorganic matter. Archetypes are objective psychoid entities actively directing and organizing ideas in the psyche and translating those organizing principles to create matter. Notice I say *psychoid,* not psychic. Psychoid means a bridge between the psychic and material.

Jung claimed that archetypes work through the psychic and material realms of existence, but also transcend them. Jung's concept of archetypes directly corresponds to the Gnostic Aeons, as centers of intelligent archetypal energy exerting their effects through various dimensions of existence. Like the Gnostics and Plato's allegory of the cave, Jung said that our conscious minds cannot directly know archetypes, we can only indirectly sense them through their effects. We perceive them as shadow copies or approximate representations.

Unus Mundus

Unus Mundus means one world, or unifying world. Jung envisioned the psychoid as the meeting ground of spirit, psyche, and matter, the place from which directed psychic energy emerges to manifest all material things.

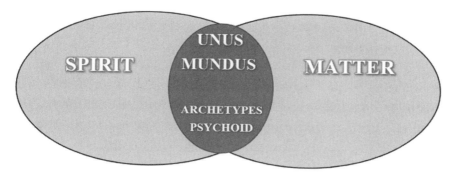

Figure 26. Unus Mundus

Archetypes organize not only ideas in the psyche, but also the fundamental principles of matter and energy in the physical world. Unus Mundus, in quantum terms, is the potential world before physical creation, and Jung's quantum-like definition caught a famous eye.

Jung's concept captured the attention of noted physicist Wolfgang Pauli who, like Jung, had vivid dreams about life's mysteries. Pauli, like other quantum physicists of his day, struggled with the puzzle of how waves of energy potential seemed to pop into existence as the particles that make up our physical reality. Specifically, how was it that human laboratory observations seemed to collapse wave energy into particles?

Pauli increasingly turned to parapsychology and psychology for answers. He started conversations with Jung that helped to flesh out the idea of this dimension from which physical reality is formed, a concept in complete agreement with the Gnostic Creation story, as we shall see.

Ego

This aspect of the conscious mind, Jung says, is devoted to the survival of the organism or consciousness. This corresponds with the Archon figure of Gnostic myths.

The Shadow

The shadow is associated with the personal and collective unconscious mind. It is a murky place where conflicting forces can create chaos. In the shadow dwells impulses and desires considered negative or undesirable that, when left repressed, can cause loss of control resulting in uncharacteristic behaviors. In extremes, it can take the appearance of spiritual possession.

However, Jung also believed that the shadow plays a positive role in balancing out our personalities. Without the shadow, the personality can become shallow and overly concerned with conforming to the opinions of others, thus stunting our growth. The trick is to integrate and balance the shadow with the personality, as light balances darkness.

Now that we understand some of the tools Jung used to probe and map the psychic world, we'll review the Gnostic Creation myth in terms of Jungian depth psychology.

17

From Mind to Matter

Matter Formed from the Psychic Dimension

DECODING THE MYTH

The Gnostic Creation myth begins with the One, the unknowable Source of all things. Both Jung and the Gnostics agree that the ground of all being demonstrates a polarity, a unity of opposites that we call by many names—male/female, positive/negative, 1/0, yin/yang—and that this polarity permeates all dimensions of Creation.

At the psychological level, we have the anima and animus, the female and male tendencies that manifest as right- and left-brained attributes. We also have the conscious vs. the unconscious. On the physical level, we have the laws of attraction and repulsion, as displayed in the structure of the atom, with positive and negative charges. Spiritually, the sacred traditions speak of light and darkness, good and evil.

Now let's take the rest of the Gnostic myth, as described in chapters 3–5, point by point. Each Gnostic event is interpreted in Jungian terms.

**The passive God projects Itself out, to form
other conscious entities.**

In Jungian psychological terms, this is the *primordial unconscious* working its way into consciousness.

**The new entities (Aeons) are centers of intelligence
vibrating with energetic frequencies representing ideals
or patterns of thought, such as love and wisdom.**

The Aeons are Jung's *archetypal patterns* or symbols, which flow down from the psychic and material dimensions and guide human thought and behavior.

Sophia breaks from the Aeons to create on her own.

This act is Jung's *individuation*—the breaking away from the blind obedience of the collective to form an individual identity.

Sophia is trapped in chaos (the Shadow).

Just as humans have an unconscious mind, chaos is Source's unconscious—the place of potential not yet given form. Virgin territory, chaos becomes the place where a new dimension can be birthed—the concept of separation and the exercise of individual will. The proto-matter that engulfs Sophia symbolizes what Jung would call the seeds of *self-realization, the awakening personal consciousness, and an act of the psychoid function that gives rise to materiality.*

The shadow (Authades) lures Sophia into chaos.

Both Jung and the Gnostics view the *shadow* as a force of opposition, and both recognize its dual purpose. In the extreme it can be harmful, but it did present the choice necessary for the spirits to exercise individual free will in the texts. Jung essentially agreed, declaring that the shadow can deter a person from shallow conformity to the collective.

Sophia births the Archon (Demiurge) to rule the new dimension.

The Archon clearly symbolizes the *ego,* the new consciousness formed to cope with and manipulate the new lower dimension. It is arrogant in its ignorance of higher forces, yet full of pride in itself as it seeks to control the environment around it. Yet, as Jung recognized, the egoic force serves as the individual survival mechanism in the face of otherwise chaotic surroundings.

The Archon creates lower Archon beings as imperfect shadow imitations of the higher order or vibration of the heavenly realm. These lower Archons, or centers of consciousness, were negative archetypes, with names such as envy and hate.

This arrangement of lower thought forms is analogous to Jung's view of the psychic dimension as the source of the shadow in our collective unconscious minds. The new tier of lower dimensions is the source of the *negative archetypes* in our collective unconscious minds that conflict with the original, higher archetypes of Source.

Sophia repents and remains present in the lower planes to contest the Archon and redeem the fallen souls.

Sophia, the continued presence of divine inspiration in the lower psychic planes, is evidenced by Jung's statement that the unconscious mind is not only the repository of the shadow but also of our *higher inspirational and creative impulses*.

The material world is created from the archetypal energy of the psychic plane that stands as the bridge between the spiritual and material dimensions.

This is exactly what Jung declared in his letters to Pauli, with his concept of *Unus Mundus*, as pictured in figure 26. "Between the unknown essences of spirit and matter stands the reality of the psychic—psychic reality, the only reality we can experience immediately."[1]

The physical world was created by lower-level (psychic) entities called rulers (Archons) or powers (St. Paul). These are actually parasitic thoughtforms granted substance by our own soul in ignorance. We created our own icons of false gods, then gave them power and control over us.

Jung says, "The world powers that rule over all mankind, for good or ill, are unconscious psychic factors. . . . We are steeped in a world that was created by our own psyche."[2]

The goal and purpose of life are to reunite our male and female aspects in order to regain the peace and love that comes with awakening to the knowledge of our higher self, our divine spark, and to transcend the limitations of physical existence.

Jung: "The soul attains, as it were, its missing half, it achieves wholeness. . . . What happens . . . is so unspeakably glorious that our imagination and our feelings do not suffice to form even an approximate conception of it."[3]

SUMMARY: THE GRAND STORY OF CREATION

Let's take the points above and summarize them into a continuous Gnostic/psychological narrative.

The Supreme being, seeking to become conscious of Itself, stirs from the Great Unconscious by the interaction and movement between Its male-female polar forces. To be conscious requires something else to observe and be observed by, so It projects other points of consciousness, which, like many mirrors, It uses to reflect back upon Itself. This collective state of spiritual unity was called the Fullness (Pleroma) by the Gnostics. Jung described this state as integration.

These points of new individual consciousness, called Aeons (Eternities), embodied high archetypal vibrational qualities or ideals such as love, wisdom, and reason. They were the pure forces or patterns that made for a harmonious existence. To possess true individuality, these new entities had to have free will, meaning choice, which in turn necessitated the presence of an opposing idea.

The archetype of God called Wisdom (Gk. "Sophia") sought a new creative experience, for it is the nature of wisdom to grow through experience. Where the other Aeonic archetypes used their free will to remain in harmony with the Whole, Wisdom chose to break with Source and pursue the idea of individual experience. Jung described this process in a key concept he named individuation.

Sophia plunged into the shadow zone of chaos, the area of Universal Mind left untouched by God. It was, in effect, the divine unconscious, the place of unrealized potential. In chaos, the high vibrating frequency of wisdom's energy mingled with a dormant substance called proto-matter (Gk. "before matter"). Wisdom's high vibration lowered upon contact with this substance, but raised the proto-matter's frequency and activated it.

A new dimension of subtle matter appeared from the energy inter-actions. Its lower frequency could not be contained in the higher vibrat-ing dimensions, so it was expelled in a violent upheaval, to become a dimension of mind-forms fueled by the egocentric force seeking indi-vidual self-satisfaction. This was the psychic dimension of mind, and all spirits that fell into it became souls, that is, spirits fallen into the illusion of separation from Source. It was a mixed place of light and shadow analogous to the collective unconscious mind (Jung's term) in human experience.

The impetus of this ego-driven force of separation continued to form a lower creation. It did so in a lower-frequency mimicry of the higher dimensional pattern of which it only contained glimpses or shadow memories. It created an abode of lower, baser archetypes min-gled with echoes of the higher, nobler images from above. This is cor-related to psychic or astral planes in popular parlance.

Jung says this tug-of-war between light and shadow was replicated, in the function of our collective unconscious minds, as the forces of limited consciousness pushed their way into physical form. The nega-tive, parasitic forces vied with the positive, creative impulses. However, just as the force of opposition was allowed to operate at the highest lev-els, the balancing force of Wholeness operated as a safety valve at the chaotic lower levels.

So the Creation became a plane of pure, platonic forms and ideals; a mixed intermediary plane of lower and higher vibrations; and a still lower and more contentious plane of physical forms. At this final level, the struggle between ignorance and awareness plays out in the minds of humans as they rise from the unconscious to the conscious.

The end game of the Gnostics is to awaken to a state where we know ourselves to be ourselves, but also know we are of God. Jung agrees.

> Our aim is to create a wider personality whose centre of gravity does not necessarily coincide with the ego, [but rather] in the hypothetical point between conscious and unconscious.[4]

Whether Jung updated Gnosticism into the modern vehicle of psychology or merely found validation of an independently developed theory in the worn papyrus scrolls is unclear. It's also not important.

Though some Gnostic material was available to Jung in the early twentieth century, the largest cache of gospels by far was found in 1945, the last year of the greatest human debacle the world has ever seen. During this time, the collective unconscious and its shadow ran amok. It was also the year that ushered in the dread era of the nuclear age.

Was it mere coincidence and serendipity, or divine providence that these silenced voices from the past were heard and revived by one brilliant man who appeared at that same turbulent moment in history, like a guide appearing in the wilderness?

Enlightening the world comes in baby steps. Jung and the Gnostics did not end all wars, but they collectively gave us the knowledge and tools that lit the way toward recognizing the angels and demons within the human soul. One day we might use their insights to prevent such global madness from again gaining a foothold.

Science always seeks independent third-party validation for proof of theories. Who would have thought that validation was lying buried for millennia in the sands of the Egyptian desert? If Gnosticism is the river of knowledge, we've just explored its tributary of depth psychology.

Next, we'll see how quantum physics adds its mighty stream to the flow of wisdom that helps us to fathom the Creation.

18

Quantum Weirdness

Bizarre Properties of the Quantum World

STRANGE VOYAGE

Imagine you are an astronaut assigned to travel to a newly discovered alien planet. You experience a vivid dream en route, where you see round balls of glowing light popping in and out of your vision, and you don't have a clue what this means.

Finally, you reach your destination, and cautiously venture out from your spaceship to examine the new world around you. You spend some time walking around, sensing something but not seeing anything. Suddenly, in a burst of light, a spherical object about the size of a beach ball appears before you.

The translucent sphere appears to be rippling with energy. You freeze, wondering how this could be—you experienced a vision of similar round energy objects in your dream when your voyage began. Your vision of these spheres has been so consistent, you almost expected to see them materialize when you landed—though it seemed ridiculous.

Soon, in the distance, you see several other light flashes as more energy spheres appear across the landscape. They are anywhere from thirty feet to a football field's length away from you. You wonder if you're hallucinating. You close your eyes, but then a burst of light penetrates your eyelids.

Your eyes fly open, and you see waves of light in front of you, but they immediately pop back into the form of the spherical energy objects. Several times, you repeat the act of closing your eyes and

125

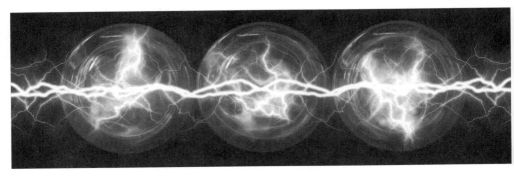

Figure 27. Wave/Particle Duality

suddenly opening them again. Each time it's the same—a glimpse of light energy waves that collapse into material spherical forms.

You now sense a weird pattern. When you're not looking, the natural state of this world seems to consist of light energy waves. But when your eyes open to observe your surroundings, the light waves collapse to a specific point, and materialize as spherical objects that look like giant atoms.

You take a step forward to reach out and touch the nearest sphere, as in greeting. The floating sphere immediately jumps back a few feet. What's more surprising is that every other sphere across the landscape jumps back at the exact same time.

Now you're really curious. You take several steps to the right. The sphere in front of you moves to the right, tracking your motion—as do all the other spheres, in split-second synchronization. The pattern repeats itself like an immaculately choreographed dance—with each movement the closest sphere makes, the others mimic it.

Then you decide to play a trick. You pull out a blanket from your backpack and hold it up like a drape, so you're visible to only the single sphere a few steps in front of you. You bob your head from side to side. Sure enough, the nearby sphere darts to and fro in short bursts from side to side.

But wait. All the other spheres still move in unison! But there is no way they could have seen your head motions. It's as if the nearest sphere instantly relayed the bobbing head information to the others.

You now draw some conclusions. First, this world seems to exist in two states—light energy and more solid particle-like forms, as evidenced by the floating spheres.

When unobserved, the natural state of the world appears as light wave energy. But when you peek, the light collapses into the more particle-like spherical objects. Then the thought hits you—could it be that *you, your mind, and your consciousness* are causing this transition from energy to particle, just by the act of your conscious observation?

After all, you had been thinking of solid spherical objects from the beginning of the voyage, as if expecting to find them. Are these spheres materializing from the light energy according to your expected observations? What kind of objects behave like that? You begin to sort it out:

- This world you've discovered seems to exist in dual states, both as waves of light energy and as more solid, visible particle forms.
- The light energy field acts like an energy matrix from which the visible objects arise.
- The material spheres seem to be particle forms of light energy, another way of saying that matter is a form of light.
- Your consciousness as an observer seems to influence the transition from the light energy state to the more solid particle forms you're observing, as though the presence of your consciousness somehow draws the energy down to mold matter from it.
- If this is the case, then the light energy field must possess a form of conscious intelligence, since it is aware of your thoughts and responsive to your presence in specific ways.
- The way they react in sync, the spherical material objects must also be in instant communication with one another, despite distances. Only conscious objects can communicate and impart directive information, so intelligence must permeate everything—you, the light energy, and the spherical (particle) objects.

Now something else happens. One of the far-away spheres flickers out of sight, then, in the same instant, appears directly behind the sphere nearest to you without any intervening motion.

It did not move across the air or ground; it just disappeared and reappeared from one remote position to a new, closer position. It went from point A to point C without ever passing through point B. This now tells you that time and space here operate like nothing in the world to which you are accustomed.

All in all, it's a bizarre and counterintuitive experience that seems like a form of magic to your normal way of thinking. It's a little disorienting and you can hardly wait to get back to the familiar operations of your own normal world.

But here's the catch—*you never really left Earth*. Oh, you thought you were in outer space? Did I forget to tell you that you were a *quantum* astronaut exploring *inner space*? Sorry, my bad. But all that you experienced happened right here on Earth, in the quantum world that lies beneath our everyday reality.

All the strange phenomena described—solid particles forming from un-solid light waves, particles in instant communication over vast distances, basic reality as an intelligent matrix of light waves smeared across space, the role of consciousness in solidifying observable matter out of light—are very real phenomena at the quantum level here on Earth.

Here's one more thing I'll throw at you: The latest theories in physics, based on observations of the deep space phenomena known as black holes, seem to indicate that the solid world we perceive may actually be a giant three-dimensional holographic projection à la the holo-deck from the *Star Trek* TV series.

Can you say Bugs Bunny? It's certainly like going down a cosmic rabbit hole. We may project Bugs onto our TV screens, but is something projecting *us* onto a universal screen? Movie within a movie, program within a program, dream within a dream? Reality is stranger than you can imagine, and nothing is as it seems.

Ladies and gentlemen, welcome to the world of quantum physics—the fascinating world we will explore in this section of our book.

19

Behold the Quantum

Quantum Physics in a Nutshell

Imagine what it might have felt like to be one of the first European explorers to set foot on the shores of the New World, gazing at the vast, open tracts of land. What animals, what people, what new horizons of experience might be waiting?

Most people still don't realize that a far greater land has been discovered in modern times, one so vast that, even today, only a fraction of it has been explored, let alone conquered. The truly amazing thing is that this is the land we all came from, a hidden land under our noses, and it is about as alien as any planet light-years away from our galaxy.

Quantum physics studies the building blocks of our universe at the microscopic, unseen level. The other main branch of physics, relativistic physics, studies the operations of large, observable macroscopic objects such as stars or planets. Reconciling the activities of the observable and the microscopic worlds has been the elusive holy grail of physics. They just don't seem to jibe.

Events in the quantum world seem bizarre and counterintuitive when compared to the macrocosmic world it creates. Another puzzle exists around the question of consciousness—in classic quantum physics, human consciousness seems to play a role in the behavior of phenomena at the quantum level.

And before we can relate quantum physics to the ancient Gnostic texts, we need to understand basic quantum concepts:

Energy Is Not Continuous

The genesis of quantum science can be pinpointed to the year 1900, with the work of a brilliant German physicist named Max Planck. Everyone believed that light was a smooth, continuous phenomenon. Planck, however, discovered that the electromagnetic phenomenon of light energy is emitted in discrete units, or packets, of energy called quanta, or light energy, in the form of particles called photons. This electromagnetic light energy is the very thing that allows us to perceive the physical world. This exploration of light's properties is what kicked off the quantum age.

Building Blocks

The subatomic particles and energies that comprise the quantum world are the building blocks of the material world. Our physical world manifests through the interaction of matter and energy. As for particles, they belong to two classes: Fermions are the fundamental particles of matter. Bosons, on the other hand, are considered to be the force carriers of energy. Motion is important at the quantum level, and the various particle types are distinguished by their angular momentum or spin. Fermions have half-integer spins or revolutions (1/2, 3/2, etc.), while bosons have whole-integer spins (0, 1, 2, etc.).

The motion of photons (energy carriers) around electrons (particles) is what generates the light by which we perceive all objects in the universe. Energy and matter, however, are really different forms of the same substance. Matter is a form of energy, and the quantum line between them is interchangeable. The play between energy and matter produces the physical universe.

Particle-Wave Duality

Particles are characteristic of matter, and waves are characteristic of energy or force fields. However, in the quantum world, subatomic phenomena display characteristics of both. The pivotal experiment that demonstrated this was called the double slit. To really understand the significance of this critical experiment, some illustrations are in order.

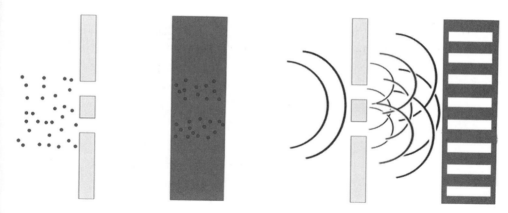

Figure 28. Double Slit Experiment

In figure 28 you see two light gray walls with two openings, or slits. The darker image to the right of each is a photographic recording plate. If we shoot marbles (particles of matter) against the wall, as in the illustration on the left, the ones that make it through the slit form a predictable pattern of two bands on the plate, in line with the openings.

If we direct energy, like radio waves, against the wall, as in the diagram to the right, the waves passing through the slits interfere with one another. Think of how ripples from throwing two stones in the water will overlap and cancel each other out at certain points of contact. The plate registers these interference patterns and the result appears as bands of light shown in the wave pattern diagram.

The behavior of the particle and waves at the macroscopic level is perfectly logical and understandable, but at the quantum level things get strange.

If we fire electrons, which are tiny bits of *matter*, against the wall, as in figure 29 on page 132, we unexpectedly find the interference pattern we would normally expect with *energy* waves. Say we reduce the flow to one electron at a time. No way they can interfere with one another, right? Wrong. Single electrons passing through the slits still produce the interference pattern of wave energy, as in figure 28.

Mathematically, the single electron starts as a particle and becomes a wave of all potentialities—it goes through two slots, one slot, and no slots at the same time. If that doesn't confuse you, try this.

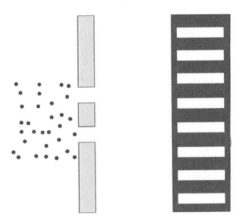

Figure 29. Unobserved Quantum Particles

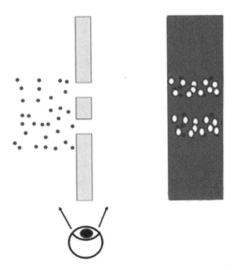

Figure 30. Observed Quantum Particles

We set up a device, represented by an eye, to observe the behavior of the electrons from the time they are fired to the time they go through the slit. Lo and behold, look what happens. The electrons now begin to behave like particles instead of waves, as we would expect. They display dual particle bands on the plate to the right. The implications from this one simple experiment have been far-reaching, and have spawned or supported numerous theories.

The act of conscious human observation appears to alter the behavior of the particles, as though they're aware of being observed. This raised questions about the role consciousness plays in creating reality.

$E=mc^2$

The double-slit experiment also affirmed Einstein's famous $E=mc^2$ formula, which states that energy and matter are the same thing. Matter comes from energy and energy resolves back into matter. Degrees of frequency and vibration determine the solidity of any object. Max Planck echoes the concept of matter created by energetic force:

> I can tell you as a result of my research about the atoms this much: There is no matter as such! All matter originates and exists only by virtue of a force which brings the particles of an atom to vibration . . .[1]

Consciousness

The double-slit experiment raised the question of how consciousness relates to the subatomic world and the creation or perception of reality. It's a valid question because our very bodies are bundles of quantum particles and energy reacting with the fabric of detected and undetected forces. The consciousness question, which we'll discuss in further detail, is a hairy one for physicists because consciousness can't be observed, at least not directly.

This is such an important topic, I devote an entire chapter to it, for reasons that will become clear.

Quantum Leaps

In the visible world, an object traveling from point A to point B must physically and visibly move through space. Not so in Quantumland. Imagine a quantum object, like an electron as a ball of light, in a dim hockey arena. You see it appear by one goal, then *poof,* it disappears only to appear at the goal on the other side of the arena. You never saw the path it took, it just appeared, disappeared, then reappeared. Now it keeps popping in and out of sight all over the arena, and no matter how hard you look, you can't see any progression trail along any pathway. No trail flashes, no tailing lights to give any indication of velocity or direction as to where it will appear next.

Nonlocality/Quantum Entanglement

This is the puzzling phenomenon of how a particle can instantly react to or know about another's state, even when separated by virtually millions of light years. It's like someone pinching you and another person on the other side of the globe instantly saying "ouch."

The effects of particles showing this entangled behavior are described as nonlocal because the affected particles are very distant from one another, and are absent any local or intervening medium. It would be like two people miles apart communicating absent phones, internet, or other electronic or mechanical means. Einstein called this unusual quantum feature "spooky."

Einstein also disliked this phenomenon because *its instantaneous communication travels faster than the speed of light,* which just ain't supposed to happen. Think about this for a second: Two far-removed particles reacting in unison aware of what the other is doing instantaneously. Awareness implies . . . you guessed it—Consciousness. Moreover, it is a *linked consciousness* involving multiple entities, not a singular self-contained event, or solipsistic phenomenon, as scientists call it.

The fact that the communication is instantaneous with no intervening transmission method has large implications. The most plausible explanation is that one is not transmitting to the other, but rather they are *both tapped into a common source or mind,* if you will, sharing common information in the manner of a hologram. The unique feature of a hologram is that any part of it contains the information of the whole.

Keep this in mind, because the holographic nature of our reality will figure prominently in our forthcoming story to give a coherent picture of how the unseen forces affecting our lives operate in seemingly magical ways.

20

I Am, Therefore I Think
Consciousness and Quantum Physics

INTERPRETING QUANTUM BEHAVIOR

The idea raised by the double slit and other experiments, that conscious observers affect quantum phenomena, is the oldest and most widely accepted view in physics. It's called the Copenhagen interpretation, after work headed by pioneer Danish physicist Niels Bohr.

This interpretation says that potential reality is *a quantum energy wave existing in all of its possible states or outcomes at once.* Only when a conscious observer measures it does one of the quantum potentials choose or collapse into an actual visible event (particles), and that collapsed state is the reality that we see.

CHOOSING AMONG THEORIES

Some competing theories have arisen over time, but the Copenhagen interpretation is the one I'll use in this book for several reasons. It has never been disproved, and the other theories all have their own problems and blank spots. The main reason I support this view comes from my own personal experiences, as described in chapter 1 and from certain laboratory experiments. Refinements of the double slit experiment (the delayed choice experiment and others) seem to reinforce the Copenhagen interpretation. I can't prove my personal experiences, but I can relate objective results from four key experiments. The first three I became aware of due to Gregg Braden's book, *The Isaiah Effect.*[1] The fourth source is from physicist Amit Goswami.

Vladimir Poponin created a vacuum in a tube, leaving only photons of light. Distributions of the photons were random. When he placed DNA in the tube, the photons formed an ordered helix pattern around the DNA helix. When he removed DNA, the photon helix remained ordered. The DNA communicated with the photons and a new, organized field of energy appeared.

The U.S. Army (INSCOM) took DNA from a human donor and placed it in a device to measure changes. The donor received emotional stimulation from a series of video clips. When the donor displayed emotional peaks and valleys, the DNA reactions had an exact correspondence. Distance made no difference. Living cells communicated through a form of energy not bound by time or space in a nonlocal manner (energy present everywhere, all of the time, rather than created and bound to one place).

Glen Rein and Nolin McCraty released a paper in 1993 that stated that within the electromagnetic field of the human heart, there is another energy. Human DNA was isolated and placed in trays held by people trained in emotional self-management. Positive emotions relaxed and expanded the DNA to the point where the helix strands unwound. With anger, the DNA contacted. The way DNA strands touch one another at different points determines how much of our dormant DNA switches on (2/3 is normally switched off), and this seems to be related to health. It appears that positive emotions can enhance the immune function and promote health and youth. The results of this experiment have also been demonstrated by the work of epigeneticist Bruce Lipton.

Amit Goswami revealed the results of an experiment he observed conducted by Mexican neurophysiologist Jacobo Grinberg. The experiment had two people in proximity meditate to communicate telepathically. They were separated, and each was hooked up to an electroencephalogram (EEG) to measure brain wave activity, inside magnetically impervious chambers. A series of light flashes was shown to only one subject. This stimulated a pattern of brain wave activity in the EEG. The other subject simultaneously displayed a virtually identical brainwave pattern when the two EEGs' results were subsequently superimposed. The experiment was repeated to rule out coincidence.[2]

These are all examples of nonlocal communication at work. So, we've seen linked consciousness working at the particle, cellular, and organ levels. If nonlocal linked consciousness is evident in minute structures like particles and cells, just what is the function and the untapped potential of the human brain?

21

Intersections

A Multidisciplined View of Creation

Up to this point, we've covered Gnostic Creation myths, Jungian psychology, and quantum physics. The evidence in this book—spiritual, psychological, and scientific—points to the conclusion that everything in creation derives from a common conscious intelligence that permeates all energy and matter, both organic and inorganic. Now let's tie these disciplines together to give us a map of how reality works.

BEING AND EXPERIENCING

We've learned that our visible world is not a fundamental reality. The true underlying reality is an energy field that permeates everything we perceive. This field, in Gnostic terms, is a light energy projection of Source Consciousness (it's okay to call It God). Source is a still, unchanging state of being. It's one thing to be, it's another thing to do—to do is to experience. Experience requires movement, action, intention, and creation.

Intention needs something to act upon. The first experience, or creation, was Source's thought acting upon its internal potential to project other points of consciousness. The newly projected beings have been called by various names: Spirits (generic); Aeons (Gnostic); Angels (Judeo-Christian); or Mind-Sons and -Daughters (Hindu), which we'll call Mind-Children.

This ensemble of consciousness revolving around Source was like a chorus acting together to extend the great theme of Creation. If Source and Its new co-creative mind-children partners were to be more than just static consciousness, a new vehicle or medium was necessary to get them out of their infinitely unchanging state, even if it was a kind of dream illusion. The ancient texts say that two things are necessary for Creation—consciousness (thought) and potential (energy).

To facilitate the creative abilities of Its mind-children, Source created a kind of playground, or arena in which the mind-children could manifest their dreams and desires. This was a field of wave energy potential upon which the consciousness of the mind-children could act, and it was everywhere at all times—in other words, an infinite field of energy potential.

Science calls this field the quantum foam or zero-point energy. Zero-point energy was demonstrated as a created vacuum at absolute zero, in which nothing should exist. Instead, the supposed vacuum was seething with quantum energies. What we think of as empty space, or a vacuum, doesn't exist. We now realize that the "empty space" surrounding us is full of invisible energy and matter called dark energy and dark matter. It's this field that the mind-children used to project their consciousness into virtual states of experience, such as stars, suns, dust, and animal life.

VIRTUALITY

The key word here is *virtual*. The projections of Source and the mind-children at this point were not physical objects. They were mind-constructs virtually experienced so that the mind-children never left the state of Oneness with Source. It was like dropping kids off at the arcade, where they could play all the virtual games but then go back home to their parents. This creative force-field playground of consciousness was what we call heaven. But then something happened.

THE FALL (SEPARATION)

At this point, the Gnostic story of Sophia kicks in. Remember, Sophia was the Aeon, or mind-child, who was not content with the collective experience of virtual creation. She wanted to create something personal, separate, and apart to act for the self, i.e., to act selfishly. Instead of using her free will to work in harmony with the will of the Source, she pursued her own path, exercising her individuality for her own pursuits. She broke the unity of heaven and plunged into chaos, setting off a chain of events that would lead to a new dimension, which resulted in materiality.

The story of Sophia was gospel in early Christianity for the Gnostics. However, since Gnosticism was so thoroughly suppressed, it sounds alien to the orthodox Judeo-Christian version of the Creation that we inherited.

I gave evidence earlier that Jesus taught a hidden Gnostic teaching for those deemed capable of receiving it, while for the masses he used parables. One of those parables is a famous mass-consumption version of the Sophia story, present in the New Testament. Jesus made the main character in the parable male because, although the Gnostics exalted the feminine, the masses were still influenced by patriarchy.

Luke 15:11–32 relates the parable of the Prodigal Son: A man has two sons. The younger son selfishly demands his share of the father's estate before his time is due. He sets off for a distant country where he squanders his wealth and ends up feeding pigs. Desperate and starving, he returns home, repenting to his father. "I have sinned against heaven and against you." The father treats him royally, rejoicing that "... *this son of mine was dead and is alive again; he was lost and is found.*" The older brother is angry. He tells the father he has been obedient all the while the younger one was squandering, yet he never received any such attention. The father replies that everything he has belongs to the elder son, but they had to rejoice that the lost son was found.

The prodigal son is Sophia. The wealth he selfishly took was the divine creative energy, the God-given creative power Sophia took and misused for her own selfish ends. The distant land was the chaos into which Sophia plunged, which separated him from his home and her from the Source. The son's squandered wealth is the precious divine creative energy Sophia misused to create a lower state of being. The feeding of the pigs was the proto-matter that fed on or drained off Sophia's divine energy.

The disappointed older son represents those spirits or mind-children that never abandoned God's will. The elder son's anger may be a reflection of the extra-biblical legend about the angels who resented God's love for humans because they felt God exalted them over the angels.

THE QUANTUM MORAL OF THE STORY

What do the Sophia and Prodigal Son stories tell us from a quantum perspective? As mentioned, Source had to create a medium with which It and the mind-children could project virtual realities from their fundamentally static, infinite existence. This medium is the energy field called quantum foam, or zero-point energy.

This energy field permeates all space. Think of it as the screen upon which the Source and the mind-children project their visions in a holographic virtual reality movie. We have now identified the two elements the ancient texts say are necessary for creation and manifestation: consciousness (thought of the higher intelligence) and potential (the energy field).

However, the creations of these high energies occurred in higher dimensions unseen by physical senses . . . until Sophia. The Sophia and Prodigal Son stories describe a shift of quantum energies directed at creating a new dream, a new dimension, a new reality—material existence. How did this happen? How did invisible energies become visible matter?

Most computers contain partitions, separate sections of the operating system existing in the same computer but walled off from one another, which have to be accessed by the operator using specific procedures. In order to simulate or project physical reality, which is to experience full individuality and separation from Source, a portion of the Source had to be segregated, or partitioned off.

Physical existence, like all existence, is a *state of mind*, but a particular state of mind that had to be separated from the energy-mind of the Source; it couldn't exist otherwise. Without the separation, it would be part of the mindset of the Source, which can't experience Itself as physical form. Pure spirit and matter can't coexist at the same vibrational level. They are two different frequencies. This is another way of describing the limitation of consciousness to experiencing individualized physical existence.

In truth, the psychic and physical dimensions are still part of the Source in the same way partitions are still part of the same computer operating system. They simply coexist as compartmentalized segments of the same infinite energy field.

So it was that the Gnostic text, *On the Origin of the World*, described Sophia's new dimension of Creation as a *"great disturbance"* in the heavens where *"matter was expelled"* in an upheaval to a *"place apart."* This is a two-thousand-year-old description of the Big Bang—the creation of matter from seemingly nothing—only, now we know what that nothing is, thanks to ancient wisdom and quantum discoveries.

CONSCIOUSNESS AS THE FOUNDATION OF REALITY

What can we take out of all this? States of mind, dimensions, energy fields—all of these are permutations of consciousness, so consciousness is the base condition of everything. The difference between material objects, such as a rock and a human being, is the degree to which they can tune into and process the intelligent information electromagnetically transmitted by consciousness.

For example, single cells can cluster to form specialized organs, which combine to form systems, which combine to form higher organic species. The result of these combined aggregated clusters is sophisticated structures, like brains and nervous systems, capable of processing information transmitted by a high-level consciousness. The brain then acts like an electrical receiver/transformer, stepping down universal consciousness into individualized consciousness, which becomes the intermediaries that produce physical reality.

The Master Conscious Energy we're discussing is intelligent, because it imparts consistent directive information the way DNA consistently controls the difference between replicating a cat or a dog. This intelligence is nonlocal, meaning it pervades and connects everything, regardless of space and time. A simple mathematical representation of this process might look like this:

$$\left(\text{Intelligence} \longrightarrow \frac{\text{information } \mathbf{x} \text{ energy}}{\text{frequency}} \right) = \text{manifested phenomena}$$

Intelligence is constant, and consists of electromagnetic quanta (packets) of information transmitted by light energy. It gets modified by dividing itself through varying frequencies. This division produces the variety of phenomena and objects found throughout all dimensions, both natural and invisible.

Materialists tend to look at consciousness in a myopic way, limited by the assumptions of their materialist methodology. Matter is the only thing they can detect and measure (or so they think), so somehow consciousness must have a physical basis. Any other way of looking at it is criticized as religious, philosophical, untestable, and disprovable.

I submit that this is bad, one-dimensional science. This materialist view of consciousness has failed to prove its case, is rife with paradoxes, and increasing numbers of scientists are thankfully turning away from it. When faced with deep dilemmas, we need to tackle the problems from all angles—science, philosophy, religion, intuition, metaphysics—and look for correlations and intersections connecting the dots, if you will.

This book looks at consciousness, the Creation, and human origins holistically, using all the disciplines listed above. If anyone is interested in seeing opposing materialist attempts at explaining consciousness from a quantum standpoint, the opinions of people like Max Tegmark, Roger Penrose, and others are readily available on the internet.

How we look at consciousness is critical to understanding the purpose of humanity and our role in the Creation. We need to understand that our apparent isolation, separation, and meaninglessness are only apparent, not real. We still co-create using the same virtual reality energy matrix as the higher intelligent energies, albeit on a much more limited, diluted basis. The role of human consciousness and the power of humanity to create and affect physical reality is the topic of our next chapter.

22

To Be or Not to Be

Our Main Thesis

A little-known but crucial debate is going on in certain circles. I say crucial because, if we had definitive conclusions, we would pretty much have solved the mysteries of existence. For now, the debate is largely between groups consisting of physicists, neuroscientists, psychologists, and philosophers. And by the way, you can throw astronomers and cosmologists into the mix too. With this motley group, you can probably understand why the average person hasn't got a clue what's going on—but one purpose of this book is to clue you in.

The topics being debated shake out this way: (a) Is consciousness a product of matter or is matter a product of consciousness? (b) Do material things have an objective existence, or is external reality actually dependent upon our consciousness? Both questions could be folded under a third consideration—(c) Does our world exist as we perceive it through our senses or are we living in a simulated reality?

By way of disclosure, it's not the purview of this book to present every argument or nuance from all sides. That would require another whole book, and it would probably cause you to tear out your hair. Anyone interested enough can take the information presented here and research opposing views independently. This book takes the definite position that consciousness is primary and external reality is dependent on the consciousness of the observer, that is, without our presence, no external world exists.

The reasons for my position are thus:

- The primacy of consciousness and the dependence of reality on the observer correlates with and explains my personal experiences, as I described in the first chapter. This also best explains the counterintuitive (weird) behavior and paradoxes of the quantum phenomena also discussed earlier.
- Quantum actions are energy waves of possibilities, not actual events or particles. Only consciousness seems to translate these energetic possibilities into actual physical events. The quantum energy waves of potential only collapse into visible particles *in the presence of observers,* hence, consciousness must be involved. A quantum event has no meaning unless consciousness is present to observe it.
- The **subject** (observer) is the person or thing *doing something,* and the **object** *is having something done to it.* In quantum physics, the quantum wave only becomes an object after the conscious subject observes or measures it. Without the conscious subject observing it, the quantum wave remains an infinite potential, not a finite observable reality. Physical properties cannot explain consciousness, but consciousness is needed to explain physical properties. In simple terms, the object—a measured quantum event or property, like a particle—does not observe you, you observe it. It doesn't collapse you from potentiality to reality, you collapse it. Therefore, consciousness must be fundamental, and matter emerges from it.
- The primacy of consciousness and the dependence of reality on the observer is the only view wherein the pieces of disciplines such as science, psychology, and spiritual wisdom fit together so well. This is exactly what this book is demonstrating.
- Scientific materialists subscribe to a bottom-up view of consciousness that runs like this:

elementary particles > atoms > molecules > cells >
organisms > brain> consciousness

This view is linked with the hypothesis of abiogenesis, the belief that organic life can arise from inorganic matter. The opposing top-down view that begins with consciousness looks like this:

consciousness > elementary particles > atoms >
molecules > cells > organisms > brain

- Materialists say that preexistent consciousness can't be proven, but abiogenesis needs many more improbable variables to even have a chance of creating conscious life.

- The great mathematician John von Neumann gave a solid mathematical basis, stating that some outside factor was necessary to collapse the energy field wave potential into an actuality, and he said this was the consciousness of the experimenter. He said that the quantum probabilities could not collapse themselves. Materialist arguments center on the brain as being just another collapsed probability measuring other probabilities, but this ignores the idea that the brain and consciousness are different. The brain is merely a physical receiver or transformer for an independently existing consciousness.

- Physicists Massimiliano Proietti and Alessandro Fedrizzi, of Heriot-Watt University in Edinburgh, performed an experimental permutation of the double slit experiment and its famous derivative thought experiments, such as Schrodinger's cat and Wigner's friend. The experiment upholds the theory that an observer's consciousness plays a central role in generating reality from quantum operations.

- **MIT Technology Review: A quantum experiment suggests there's no such thing as objective reality.**
  This experiment reinforced the idea that our conscious minds collapse energy into the particle forms that create the physical world. From the MIT Abstract: "The question whether these realities can be reconciled in an observer-independent way has long remained inaccessible to empirical investigation, until recent no-go-theorems . . . If one holds fast to the assumptions of locality and free-choice, this result implies *that quantum theory should be interpreted in an observer-dependent way.*"[1]

SUMMARY

The reasons I just listed form the basis of this book's thesis, which I will repeat here. When reading the thesis, refer to figure 31, in which:

- **S** represents the Source.
- The lightning bolts are electromagnetically transmitted conscious information.
- The alphanumerics **D1–D4** are various dimensions of consciousness as projected by Source. The entire circle represents the Source, the Whole, or the Oneness of all existence.

Notice how the lines delineating the dimensions get thinner the farther they are from Source. This graphic shows how the frequency or level of consciousness becomes more limited as consciousness moves further away from the vibration of the ground state we are calling Source.

This diagram not only reflects the Gnostic conception of consciousness and dimensions of existence, but it also forms the basis for many quantum theories such as parallel universes and the many-worlds theory.

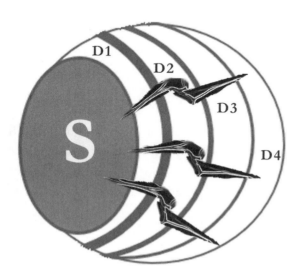

Figure 31. Dimensional Projections of Consciousness

The thesis of this book was stated in chapter 1, to which you may want to refer, but here it is again, stated in different terms in light of what you have read since:

- Objective reality does not exist as our senses perceive it. What we perceive are electromagnetically transmitted images of a Master Consciousness received by the brain and projected as three-dimensional materiality.
- This Consciousness is an intelligent energy that permeates everything from an electron to a rock to the human mind, linking all creation together. It manifests itself in varying frequencies of energy or vibration, like radio waves.
- It manifests not only as invisible and physical objects but also as differing, interpenetrating dimensions of existence. These dimensions form, affect, and support each other like stacked pancakes.
- Our personal consciousness derives from the Master Consciousness but operates in a greatly limited frequency of awareness that makes us think we're something separate from the Whole.
- Human consciousness is not just the result of the force acting through the energy field, it is a co-creator and manipulator of that field.
- Human minds are the bridge between the material world and the dimensions of higher consciousness/wave potential, but we must awaken to this reality to fully and consciously activate our innate co-creative power.

The diagram on page 147 illustrates one of many examples you'll see of how shockingly advanced was the two-thousand-year-old Gnostic wisdom. In a 2008 TV video, Michio Kaku, noted quantum physicist and co-developer of String Theory, surprises his interviewer with some heady statements.[2]

Kaku gives the example that we are surrounded and permeated by radio signals from NBC, BBC, or Radio Moscow, but we can only tune into one frequency at any given time. We are de-cohered or ignorant of the others. He then states that the universe similarly is *vibrating* (his words, not mine), and that many universes are vibrating in the very room of the interview, universes where the dinosaurs never died and where aliens are present on a dying Earth.

We can't detect them, though they surround and interpenetrate us, because we have de-cohered with them, i.e., we are not in tune with them. Our universe is tuned to one frequency only—our universe—but reality consists of many dimensions, as a multiverse. The awakened Gnostic masters who tapped into higher consciousness were able to experience or be in tune with these other dimensions, and so they were able to prove things, like Dr. Kaku's statements, thousands of years ago.

Bottom line—Gnostic wisdom and quantum physics are tied at the hip, so if you want to toss out Gnosticism as superstitious arcane nonsense, be prepared to do the same with the quantum science that has shaped our modern world.

23

Through a Glass, Darkly

The Illusory Way We Perceive the World

We've studied the proposition that objective reality does not exist and consciousness is dependent upon the observer. Let's now examine that proposition in a more personal way. It is a fact that we never directly experience any reality of "what's out there." To understand what this means, we need to look at some information that has come from neurophysiology.

HOW WE SEE

The objects we think we're perceiving, like an apple, for instance, are not something we directly experience. What we experience are indirect reflections of things in the back of our brain.

In figure 32, photons of light reflect off the apple and are picked up by the cones and rods in our eyes. The photonic images are then electrically converted into neural impulses, as depicted by the wavy white line. The impulses are transmitted and experienced as an image in the vision center at the back of our brain.

So, we never have an objective, direct experience of something called an apple in the external world. Rather, we experience the apple inside our heads as an electrical image in the dark, enclosed space of our brain. We don't really know what an external world looks like. It's more accurate to say we are seeing a neuro-electric facsimile, the same way watching a movie doesn't show the real actors but electronic images of them. What we are seeing is a second-hand view modeled along the lines of a movie we call reality.

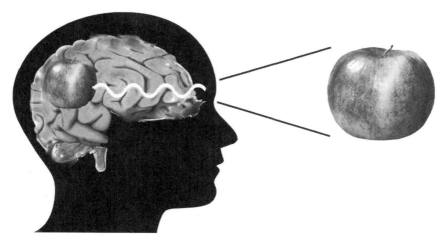

Figure 32. Reality Perceived as Light Images

WHO'S RUNNING THE PROJECTOR?

You may say, Well, even if we don't experience it directly, the apple is still out there as an independent reality because the photons of light must reflect off something to register in our brain. True, but the sum of everything we've learned from the Gnostics, physics, and psychology leads to another conclusion. Remember, atoms are 99 percent light and space. We don't see or feel objects because of the 1 percent portion of matter—we see and feel objects as projections of light energy, which is the electromagnetic charge caused by the polarity of subatomic particles moving in space.

More than likely, the "objects" are organized electromagnetic light projections, not objectively existing objects. And who's doing the projecting? The Gnostics, Jung, and some brave physicists might tell us that our basic reality projections are coming from other dimensions of archetypal intelligence. Call it the action of God, consciousness, or the cosmic rabbit, but it challenges the very notions of orthodox materialistic science.

A growing area of research called BCI (Brain-Computer Interface) deals with wiring brains directly to computers. This harkens back to the 1960s when Gilbert Harman, a philosopher in the United States, proposed what was probably, back then, a rather outlandish thought experiment.

If a mad scientist were to remove a human brain from its body, suspend it in a jar of some liquid that kept it alive, and connect its neural system to a computer, the electromagnetically transmitted images from the computer, like the apple in figure 33, would become the experienced reality of the brain. The brain would have no way of distinguishing the truth from what the computer was showing it. The brain, therefore, would be acting like the receiver we discussed before, and the computer would be its perceived consciousness.

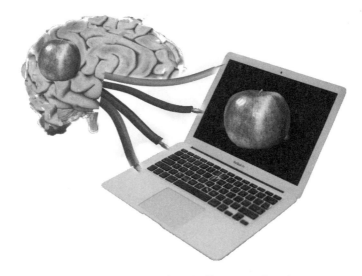

Figure 33. Reality as Electrically Transmitted Images

TIME AND MOTION

The next question we might ask is, how does the perception of motion come into play to contribute to our sense of reality? This is where time comes in, and it may have to do with the functional purpose of time itself.

Max Planck, mentioned earlier, founded quantum physics with the observation that light is not continuous, but is emitted in separate units of energy called quanta. As a result, some physicists, like Julian Barber, have perceived the underlying quantum universe as composed of a static array of separate quantum frames, like a strip of film. This was also Einstein's conclusion. The Gnostics wholeheartedly agreed when they said God is unchanging.

So, we can picture fundamental reality as a static, unchanging array of organized images, units of information, or "snapshots" as Barber called them, which nicely corresponds to the Gnostic concept of distinct archetypal images providing the basis of our reality.

But what exactly are these cosmic frames made of? The Gnostics or Carl Jung might call them archetypal images but, in terms of physics, they are units or quanta of light-encoded information called Planck Lengths or Planck Frames, after Max Planck. The theory is that our universe is "papered" or surrounded by two-dimensional sheets of these information-packed Planck Lengths (see figure 34a).

Knowing that life is motion, how do we get from an underlying static foundation to a universe where everything seems to be moving? The answer appears to be that we're experiencing a light show of cosmic proportions. Let's examine that idea with figures 34a and 34b.

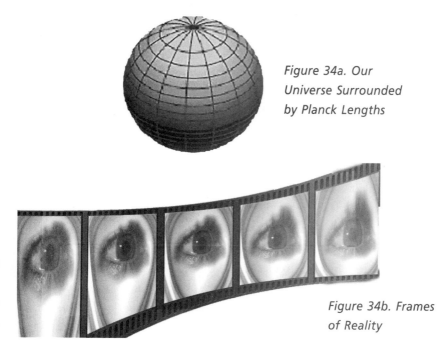

Figure 34a. Our
Universe Surrounded
by Planck Lengths

Figure 34b. Frames
of Reality

Imagine the static fundamental frames of reality we spoke of as a strip of still film frames, as seen in figure 34b. A film, or even simple flipbook animation, is a series of separate, static images, but when we project or flip them in time sequence, they take on the illusion of motion. So, even though the ground state of reality may be motionless and changeless, our sense of motion and change is caused by timed sequences of light energy information revealed like film frames passing through a projector. Time, therefore, is as necessary as light for composing the reality we perceive.

We experience the information contained within the frames or Planck units as time-sequenced, light-illuminated 3D objects. We'll discuss the idea of living in a holographic universe and what might be projecting our cosmic movie in the next chapter but, speaking of 3D, let's look into how spatial perceptions arise and how we distinguish objects.

PERCEPTION IS EVERYTHING

We've covered how we perceive objects and how those objects appear to be moving in space, but what about our actual perception of objects as things in 3D space? Like it was with motion, it comes down to light and time. Our perception of objects, remember, comes from photons of light bouncing off objects and reflecting images through our eyes to our brains. The thing is, all that light-encoded information doesn't reach us at the same time. It happens in sequential increments, as shown in figure 35.

The eye on the left is viewing three main objects—a temple, an elephant, and a woman—all of which are at varying distances. The dashed lines represent photons of light reflecting off the different objects. The farther the object, the more time it takes to register on the eye, so the woman would be detected first, followed by the elephant, and finally the temple. These time variances give us the perception of space, distance, separation, and the sense of 3D objects.

If all things were viewed at the same time, they would appear as a mish-mash of flat images. If no time was involved, we would see nothing because it takes time for light to travel to our eyes. Space without

Figure 35. Perception of 3D World as Time-Delayed Light

time cannot produce three dimensions, which is why we use the term space-time to accurately describe our universe. Physical dimensions just cannot be perceived to exist without the sequential aspect of time to delineate them.

This view of an underlying reality may seem hard to grasp, but it does indicate that a simulated reality is necessary to make life as we know it possible. I think that one day, through enlightened consciousness, we will see through the illusions more clearly, as it states in the Bible:

> For now we see through a glass, darkly; but then face to face: now I know in part; but then shall I know even as also I am known.[1]

24

Worlds of Possibilities

Incredible Theories of Quantum Physics

In the preceding chapters, we discussed basic quantum phenomena and operations. Now we need to examine quantum science's best and most intriguing theories about how existence plays out in light of quantum discoveries.

PARALLEL UNIVERSES

One of the ways quantum physics is trying to explain the operations of our world is by postulating the existence of parallel universes that stack up with ours, or even interpenetrate ours. Without going into confusing technicalities, the basic idea is that the forces that shape our world can be better understood if we hypothesize the existence of other underlying dimensions. Theories such as the Many Worlds Theory, the Multiverse Theory, String Theory, and Brane Theory are all variants of parallel dimensions. String Theory even explicitly uses the notion of vibrations or frequencies that distinguish phenomena. This is exactly in keeping with the vision of underlying dimensions described in the Gnostic texts.

HOLOGRAPHIC UNIVERSE THEORY

A hologram is a structure wherein all the information comprising it is distributed over every part of the structure. So, if you zero in on any single part, it will contain the image of the entire hologram. Eminent physicist David Bohm and psychologist Karl Pribram developed a theory that the brain operates in a holographic manner. Bohm believed that everything, even our thoughts, is just appearances.

He postulated the existence of a realm of pure information (the implicate order) from which physical, observable phenomena unfold as a combined series of limited views (represented by the frames or snapshots in the previous chapter), but we can never see the source from which materiality arises. Material objects as limited views arising from a pure, unseen hidden order—this is *the* central Gnostic Creation belief of physical reality as limitations of the One Consciousness, as described in prior chapters.

Actual evidence for a holographic universe came with the study of black holes. The surprising find was that the information in a black hole—be it particles, debris, whatever—is not stored in the volume or inside the black hole. Rather, it resides on the surface area and is projected inward. This is a bit tricky to understand, so I'll use illustrations.

THE UNIVERSAL BUCKET

Imagine, as we discussed in the previous chapter, that our finite three-dimensional universe is enclosed in a sphere covered by a series of two-dimensional sheets (see figure 36). The grid sheets represent the two-dimensional boundary or surface area of the universe so that everything inside, including the Earth, is within the volume of the sphere, like contents inside a bucket. This means all the information that makes up our three-dimensional reality should lie within the space or boundaries of our universe. Sounds logical, right?

Ah, but black holes tell us something different. Black hole

Figure 36. Our Universe, Surrounded by two-dimensional Planck Lengths

observations indicate that the Earth and all other universal features (information) don't actually originate inside the space or volume of our universal sphere but on the outside of its *surface area*. Using the bucket analogy, reality would reside on the exterior siding of the bucket, not inside it.

Research indicates there is an edge or boundary to the universe, a cosmological horizon, a place where light hasn't reached. This corresponds to the sides of the bucket in our continuing analogy. The theory is that flat, two-dimensional sheets of light energy called Planck Lengths blanket the *edges* of the universe. This is depicted in figure 37, where the universal sphere on the right is encased in Planck Length units shown as grid patterns. Planck Lengths are units of light energy that contain the information that organizes and maintains our universe.

A hologram is light from a source, reflected off of two-dimensional surfaces to derive a three-dimensional image. Many noted physicists, such as Leonard Susskind, Gerard 't Hooft, and Craig Hogan argue that the entire universe operates according to such a holographic process, not just black holes. Figure 37 shows how a hologram might work on a universal scale. The labels on top are Gnostic conceptions, and those on the bottom are quantum terminology.

A source, or God (the projector), projects light energy as information onto a flat screen composed of two-dimensional, frame-like Planck Lengths, analogous to spiritual archetypes. These Planck Lengths are like a quilt, forming the "skin" or boundary of our universe. They filter light energy, contributing their unique qualities to form the totality inside the 3D space we perceive as our universal reality. Like a movie on a screen, the collective information is projected from without to create a holographic simulation of our universe within, but the process is not mechanical. Rather, it's a multifaceted electromagnetic energy display, if you will.

The question mark on the projector in figure 37 signifies that science has no idea what the source is. In fact, science has no idea what any form of energy is, in and of itself. They can manipulate it and describe its effects, but they have no idea where energy comes from. However, the Gnostics think they know. Looking at the labels across the top of

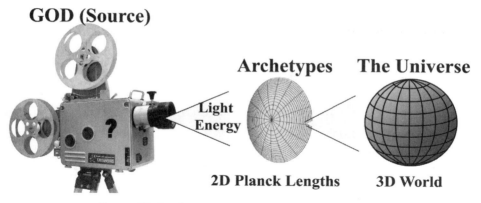

GOD (Source)

Archetypes   The Universe

Light Energy

2D Planck Lengths   3D World

Figure 37. Reality as a Projection of Light Energy

the diagram, God (Consciousness) is the light source projecting the archetypal thought images that form the various informational grids to impart specific directions to our phenomenal world.

If this sounds preposterous, do a YouTube search of "3D building facade projections." You'll be blown away by holographic projections of entire buildings right on actual city streets! Remember, "as it is above, so it is below." The very fact that we can create such wonders comes from deeper patterns operating in our underlying reality. Our scientific and technological progress merely mimics or remembers things already present in the other dimensions from which our consciousness flows.

Like every idea in science, common information will give rise to conflicting interpretations—some of which oppose the holographic theory. Just as I said earlier, with the Copenhagen interpretation, I look for correlated data to validate, and make an interdisciplinary judgment call to form my theses.

The holographic nature of reality makes eminent sense out of so many mysteries. For instance, the connections we saw between distant particles, remote DNA interactions, and people displaying faster-than-light communication with each other make sense only if they are parts of a hologram where each part possesses information available from the whole.

The central Gnostic premise of a conscious force that projects the image templates (archetypes) that ultimately form the 3D world fits like a glove with the holographic universe concept, as does the Jung-Pauli archetype concept of Unus Mundus or the unified world. The belief that the universe has a cosmological boundary, or edge, also correlates with the Gnostic concept of boundaries or veils that separate dimensions. The M-theory, or brane theory, in quantum physics reflects the idea that a kind of membrane is present, separating dimensions. The idea of boundaries is also the basis for the computer partition analogy mentioned in chapter 21.

Quantum mechanics has shown that we see and feel objects because of electrical charge, not because of the properties of mass. Remember, the atom is only 1 percent matter and 99 percent space and light. Electric charges result from the action of light, and what is light but a medium for projecting? The Creation is much more a question of light than matter, because even matter ultimately is just light energy in lower frequency forms (remember $E=mc^2$?). Again, weren't we always told that God is light?

When you consider this and the information that follows, it's not difficult to conclude that we are living in a giant holographic projection, a cosmic play of information-containing light energy. Let me take some quotes from an article by Marina Jones, on the great physicist David Bohm, a protégé of Einstein's and a proponent of a holographic universe.

> Bohm believed that his body was a microcosm of the macrocosm, and that the universe was a mystical place where past, present, and future coexisted. . . . Bohm considers *our separateness an illusion* and argues that at a deeper level of reality, we, as well as all the particles that make up all matter, are one and indivisible. . . . If more scientists like Bohm were willing to treat physics as a quantum organism rather than as quantum mechanics, we might get closer to a revolution in our understanding of the universe.[1]

Hop on Board

People as diverse as scientists of all branches to Elon Musk are giving credence to the holographic or simulated reality concept. Look at the

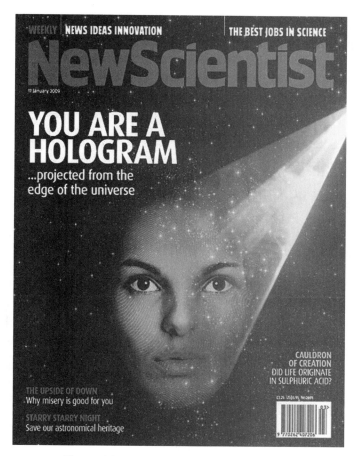

Figure 38. New Scientist *magazine cover:*
YOU ARE A HOLOGRAM

cover of *New Scientist* magazine in figure 38, which proclaims: "YOU ARE A HOLOGRAM . . . projected from the edge of the universe."

In the next chapter, we'll take a closer look at the energy field or screen from which the cosmic hologram is produced.

25
Swimming in Energy
The Hidden Matrix of Creation

QUANTUM FIELD THEORY

By now you may have gathered that the quest to find and examine subatomic particles, on which quantum physics has been fixated since its beginning, is important but not exactly conclusive in the mission to uncover what life is about. The reason? Well, pretty much every quantum scientist agrees that the real foundation of existence lies with energy. The deeper they dig into particles of matter, the more this becomes obvious.

Austrian physicist Walter Thirring articulated it this way:

[modern physics] . . . has taken our gaze from the visible—the particles—to the underlying entity, the field. The presence of matter is merely a disturbance of the perfect state of the field at that place; something accidental, one could almost say, a "blemish" . . . Order and symmetry must be sought in the underlying field.[1]

Tim James, writing in the November 6, 2019, edition of *BBC Science Focus Magazine,* gives a good summation of field theory. He says to scrap the notions of particles altogether because *they don't really exist.* He uses the images of a tornado moving through the air and a wave rolling across the ocean. They both appear to be self-contained objects or things in their own right. However, they're actually just fluctuations in the normally smooth basic states of air and water.

Accordingly, field theory variously describes particles as fluctuations, blips, local disturbances, or blemishes in the quantum energy field. As such, James says they are a form of illusion since they are not fundamental but rather like temporary blips on an otherwise permanent screen. So, *The Universe is a collection of overlapping fields and what we think of as matter is just three-dimensional bumps forming in these fields.*

NEWSFLASH
Brookhaven and Fermilab Muon g-2 experiments

As I write this book, an article appeared in *SciTechDaily*. Scientists have been frustrated by the knowledge that the Standard Model of physics, in use for decades, cannot possibly account for all phenomena, such as dark matter. Recently, they conducted some groundbreaking experiments. Without getting too technical, they worked with exotic particles called muons to achieve deeper forms of measurement, producing unexpected results. The scientists believe they're on the verge of discovering a deeper layer of reality. See the quote below, summarizing their conclusions:

> The answer is that space itself is not empty; what we think of as a vacuum contains the possibility of the creation of elementary particles, given enough energy. In fact, these potential particles are impatient and are virtually excited, sparking in space for unimaginably short moments in time. And as fleeting as it is, this sparking is "sensed" by a muon, and it subtly affects the muon's properties. Thus, the muon magnetic anomaly provides a sensitive probe of the subatomic contents of the vacuum.[2]

This single quote supports everything I have stated so far—space is not a vacuum. An energy field of virtual particles exists, a place from which elementary particles (blips in the field) can be created. This field can "bleed through" and affect substances in the visible world, such as muons. It all harkens back to Sophia activating proto-matter in the virtual potential of chaos.

The evidence of quantum blips forming reality seamlessly corresponds with the snapshots or frames shown in previous diagrams. The "reality" frames or "quantum blips" are virtual particles turned physical. Within the smooth flow of the energy field, they are like partitions or outcroppings containing physical reality. Think of your typical space movie—you have the main ship and you have escape pods attached to the main ship. The pods are part of the whole ship, but they are also compartmentalized partitions that are a space of their own, just like the computer partition analogy we used earlier. Both examples describe how our blip of physical reality can be part of a greater Whole without seeming to be so.

Remember how the Gnostic texts described the first appearance of matter as a "great disturbance" partitioned or set apart from the smooth perfection of the heavenly Pleroma, or Fullness? You might think all these Gnostic insights, so similar to psychology and physics, are mere coincidences, but they are surely starting to pile up. At some point, you have to cross the line from circumstantial coincidence to meaningful insights, so let me give you another quantum example.

DARK MATTER/DARK ENERGY

Quantum and astrophysicists have strongly postulated the existence of two invisible substances in our universe called dark matter and dark energy. These substances don't interact with light, and so, are invisible to us, hence the term "dark." Here are some surprising statistics: Dark energy is estimated to comprise 68 percent of our universe, and dark matter 27 percent. That means less than 5 percent of the universe is made of baryonic or visible matter. How'd that happen, and what is this dark stuff anyway?

Several strong clues regarding the presence of dark matter exist, but the most telling is that there isn't enough regular matter to keep the galaxies in the universe together by gravity from visible matter alone. For clusters of matter to clump together and form galaxies, a far greater attractive force has to be present and that, it's theorized, comes from a massive amount of invisible matter filling up space. This is called dark matter.

But scientists were surprised to discover in the last few decades that the universe as a whole is still expanding, and doing so rapidly. The only thing that can account for this movement is the presence of an invisible energy force pushing that expansion along. That is dark energy.

Why are these phenomena invisible? Again, they must be composed of energy or particles that don't interact with light or other particles, thus rendering them undetectable. Some candidates for dark matter substances are WIMPs (weakly interacting massive particles) and axions. Speculation about where these substances originate from is all over the board and I could not possibly cover them all here. However, I'm not alone in the opinion that these forces were present before the Big Bang, and that *they originate or bleed through from other dimensions.*

Why do I say this? A majority of the physics community believes in one form or another of parallel universes. They do so because it helps explain phenomena in our universe. Why can't the dark forces be explained as originating from parallel or underlying dimensions? One view holds that dark energy consists of virtual particles continually popping in and out of existence—so where do they pop out *from*? Other dimensions are good candidates.

These ideas are very reminiscent of the Gnostic Creation story. The proto-matter into which Sophia fell sounds very much like virtual particles, or matter-in-potential. When Sophia's energy activated them into actual matter, the implication in the texts is clear—it is a finer, more subtle matter making up a psychic dimension, not the denser matter of the physical world.

Gnostic and related ancient wisdom envisioned layers of dimensions, matter, and forms. Ancient traditions, such as the Hindu and Chinese, recognized that humans are multidimensional beings with aspects of ourselves present and corresponding to the forms of each dimension. These dimensional layers range from highly vibrating matter to the subtle material matter comprising the invisible psychic body, and finally to the dense matter of sensually visible forms.

Thus, energy is manifested in stages of ever-densifying matter, going down the dimensional ladder to the densest physical dimension. Lest we get too New Age, at least one physicist, Deno Kazanis, has thoughts along similar lines:

We can no longer assume that all matter was created at the Big Bang, even though luminous matter clearly was. Some form of "dark matter" may have existed before the Big Bang. . . . Instead of the universe going from no time-space-matter to our present sense of time-space-matter, the universe was created in different stages of time-space-matter. . . . If humankind was indeed composed of subtle bodies [as ancient texts claim] made up of different types of subtle "dark matter," and if we were capable of "shifting" our consciousness to these other subtle bodies, then we would indeed experience the world of that type of matter, much like mysticism states.[3]

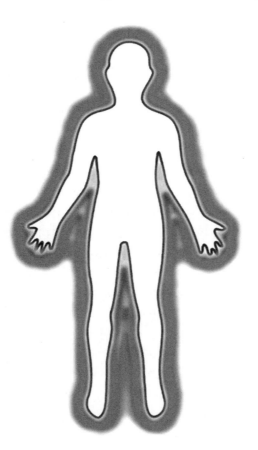

Figure 39. Subtle Bodies

Are dark matter and dark energy links to the realms the Gnostics say preceded ours in Creation? Possibly. Given the descriptions of the physical, psychic, and spiritual planes we have seen so far, it's as good a hypothesis as any.

INFORMATION THEORY

Classic information theory, developed by Claude Shannon in 1948, had to do with the mathematical coding and transmission of information by mechanical means. Information, as we are discussing it, has analogs to the classical conception, but here we're referring to information on a quantum rather than a mechanical basis.

Information is not abstract. It has both energetic and physical components. It is always physically conveyed, another way of saying that information has detectable mass. Information is not the mere transmission of data through books or verbal exchange, though even these, too, have a physical basis. The information you are reading here was stored and reproduced through binary electronic computer bits that have particle mass. The ink and paper of books all contain subatomic particles, without which there would be no means to reproduce and transmit anything by writing.

The thoughts and speech a lecturing college professor uses to impart data are stored in his brain and transmitted by electrochemical carriers running through a neural network. This is a physical process. The ubiquitous internet is a great web of fiber-optic and other links. The information it delivers to you electronically comes through packets of photons in binary (1/0) strings running through fiber-optic cables. You get the point. Information is always conveyed by a physical medium, no matter how small, invisible, or seemingly abstract.

Photons, or particles of light, are information-bearers. This is a fact. We can program, store, and transmit information ourselves using photons in binary code, which is the basis for all computers. Photons with right-spins represent 1 and photons with left-spins represent 0. These are called "bits" of information. By sending a beam of variously arranged photon particles through a transistor in strings of 1s or 0s, we are physically conveying bits or units of information. This is similar to how separate DNA nucleotide components impart information to the human body by arranging themselves in different sequences in the DNA helix.

The next revolution in computing, quantum computers, is well underway to reaching beyond duality and proving the infinite unity of all things. Instead of either the 1 or 0 bits used in ordinary computing, quantum computing uses underlying quantum substances to compute information in multiple states simultaneously. It's not 1 *or* 0, it's 1 *and* 0.

The information is not stored in this place or that, but in all places at the same time, in a quantum phenomenon called superposition. Superposition allows for all things to occupy the same space at the same time, something seemingly impossible in our everyday macroscopic world.

These superpositioned units are called *qubits* of information, as opposed to conventional bits of information. Qubits will make it possible to compute information at exponentially faster speeds, as well as tackle problems too complex for the current generation of conventional computers.

Remember this from our book's central thesis—*Consciousness is an intelligent energy that permeates everything, from an electron to a rock to the human mind.* The new generation of computers is tracking the workings of the Infinite Mind by using Its ability to be everywhere all at the same time. *As it is above, so it is below.*

GOD IS LIGHT

We've learned that without light we could not detect physical reality, since it is the interaction of light photons that makes everything visible and touchable. So, light energy "in-forms" us by presenting the world of created forms. We've learned that photons of light, like other particles, can be in communication with one another instantaneously, from great distances.

We've learned that photons respond to human DNA (see chapter 18), and that they carry directive information, which is intelligence—which strongly implies consciousness. Now science is learning to manipulate the infinite properties of light to pierce veils of knowledge previously off-limits to us. Is it possible that science is now approaching the realm of the infinite the Gnostics touched upon by using consciousness to peer into the same dimension?

The Gnostics and other spiritual and religious traditions have pro-claimed over thousands of years that God is light. Do you think, per-haps, they were onto something all along?

YOU BE THE JUDGE

Materialist scientists will criticize some of the contentions stated here for being untestable, philosophical, or—even worse—spiritual. I respond by saying that one day they will have to remove their blinders. It will take more than science alone to get to the root of reality. At one point, many currently accepted theories appeared untestable. "Baseless" intuition, such as Einstein's initial insights, led to remarkable proofs, and what is untestable today may be testable tomorrow. But if we stifle reasonable hypotheses today and they are labeled as blasphemy by the orthodox scientist-priests of the modern age, we may never even get to the great theories of tomorrow.

To say that between science and metaphysics, never the twain shall meet puts science into the same dogmatic attitude of the religion so many scientists look down upon. Orthodoxy of any kind stifles growth. At the very least, science should take clues from other disciplines, such as spirituality, metaphysics, and meditation, as starting points for novel inquiries, which might one day lead to scientifically testable theories.

I believe we possess all of life's answers within ourselves. Learning is more about unlocking the knowledge within than acquiring knowl-edge from without. Mysticism and metaphysics can be great aids in this quest. Books like this one provide signposts, but they're only of real value if the content resonates with what feels true to you, the reader. Science has a way of shooting down intuition. Don't let yourself or your ideas be stifled by anyone in your search for life's meaning. Experience has a way of correcting you, if you're off track. Listen to all sides but, at the end of the day, be your own judge.

26
Consciousness, the Hard Problem
How Consciousness Creates Reality

LOOK WHAT'S IN THE ROOM

Most physicists prefer to avoid the question of consciousness but, like the 800-lb. gorilla in the room, they just can't. Physics falls into two general categories—applied and theoretical. The applied guys couldn't care less about consciousness, the source of energy, or any questions of origins for that matter. Physics is like a kid's toy to them. As long as they can manipulate it to perform tasks, they don't care who manufactured it. That's just not what they're paid for.

The theoretical folks come from a different orientation. As their name indicates, they come up with theories about the deeper nature and operations of quantum phenomena. They, however, are still limited by the view that nothing is real or worthwhile unless it is testable and disprovable. I pointed out that this is fine in a narrow framework, but narrow frameworks won't cut it when attempting to understand things beyond space and time—and that 800-lb. gorilla certainly transcends space and time. That's why, in the world of physics, they call consciousness "the hard problem."

TALK TO THE BOSS OR THE SERVANT?

What makes the problem so hard is that the materialist ceiling under which scientists operate doesn't allow for transcendent possibilities that are considered metaphysical or, even worse, religious. So, they keep looking to find consciousness within different particles, thinking consciousness is just some firing of nerve synapses in the brain. How on earth does that account for the human ability to *feel* things, as opposed to simply perceiving objects?

Sure, you can see a flower and account for that observation via electrochemical brain reactions, but to perceive its beauty, to feel compelled to capture it artistically or express it poetically, is of a higher order entirely. If this wasn't the case, we'd be more like robots and less like humans. We identify with ideals or archetypes that trigger deep inner emotions. Attempts to explain these things by materialist mechanisms seem woefully inadequate, as do explaining all the things we've cited, such as telepathic communication between particles, cells, and people.

Now, it's certainly possible, maybe even likely, that one day we'll find physical links with consciousness, that is, substances that act as relays or carriers from immaterial consciousness to the physical domain. The brain itself is a receiver organ, but what subtler physical substance might carry information to the brain? Roger Penrose looks at things like microtubules to answer this question. Other people are looking at interesting theories like axions in dark matter as consciousness relays that feed our cognitive processes.

Consciousness lies in the quantum field permeating all space. Identifying the carrier of consciousness to the human brain would be fantastic. But, even if we find such a particle or substance, we can't mistake the messenger for the one sending the message.

If consciousness has a physical carrier, finding it is a great event, but it's kind of like talking to the servant instead of the big boss. The servant can provide information about the boss, but it is still lower down in the chain of command, and we're looking to get as deep into the origin of conscious life as we can.

I somehow feel that, if a carrier medium was ever found, many scientists would jump to tout it as solving the entire mystery of consciousness, instead of viewing it as another clue to the ultimate puzzle. James Herrick, Professor of Communications at Hopewell College, puts a proper perspective on the issue. He distinguishes the physical linkage question from the source. He hits upon the real prize, which is discerning the ground state of existence:

There is . . . quite possibly a common physics linking human consciousness to the ground state of "everything that ever existed or can exist" in the universe [and we] are part of it. Each one of us as an individual is an excitation of the vacuum [blip], an individual being on the sea of Being and this is a straightforward conclusion of "orthodox physics." It is . . . proven.[1]

DIFFERENT STROKES FOR DIFFERENT FOLKS

We've studied the nuts and bolts of consciousness in previous chapters, so now let's look at the consciousness question from the high-altitude perspective of some eminent scientists. We can start with the father of quantum physics.

Max Planck

Behind the existence of all matter is a conscious, intelligent mind— this mind is the matrix of all matter.[2]

Planck's basic equation that kicked off the field of quantum physics is $E=hv$. In this equation, E is light energy, h is called Planck's constant, and v represents the individual frequencies of the base energy. In plain English, it means that the light energy (Master Consciousness) that pervades the universe is constant, but it is modulated or differentiated into separated units, or quanta, of variable frequencies.

Does this remind you of anything? Remember the Gnostic description of how God emanated and differentiated the various dimensions or entities called Aeons (Archetypes) by vibration or frequency? The ancients likened this to a master flame that would light other flames below it, but the master flame remained constant, that is, it would burn eternally. Each individual flame or candle in figure 40 was manifested by a variation in the vibration or frequency of the master flame E.

In the Gnostic texts, the intelligence and creative energy of God are always equated with light energy, pictured as the great master flame E, at the top. Applying Planck's formula $E=hv$ to the Gnostic texts, we get

Figure 40. Scientific Interpretation of Gnostic Emanation

this: E is the eternal divine light energy that permeates Creation. The h indicates that this energy is constant, unchanging, and permeates everything. The v is the frequency or vibrational variation of the constant that imparts individual characteristics or identity to the base energy. This results in beings that are individual, yet share in the energy of the source that projected them.

Restated for clarity, God's constant, unchanging energy lights the candle below it (new consciousness) by lowering its frequency ($h \times v$), then that candle lights the next one down, and so on. The lower the candle, the more the consciousness diminishes. However, during this process of generating individual dimensions and consciousness by light transmission, God's master flame (light energy) is in no way diminished, so it is constant, as in Planck's constant. Again, it's like burning a CD, where the original remains the same, but the copies of copies each lose degrees of clarity, or fidelity.

Little did Dr. Planck realize that his formula was giving mathematical expression to the operation of the conscious light energy that created everything seen and unseen, a process identified two thousand years earlier by a group of suppressed mystics.

MODERN CONCEPTS WEIGH IN

Before we summarize conclusions from the material you've just read, let's see a random sampling of current beliefs on the nature of consciousness and reality, from scientists of differing disciplines. Naturally, you'll see some variation of opinion as well as some common threads.

Astronomers Jeans and Eddington

The two prominent astronomers James Jeans and Arthur Eddington had this to say about considering the operation of the physical universe:

> . . . the universe begins to look more like a great thought than like a great machine. *Mind no longer appears to be an accidental intruder into the realm of matter . . . we ought rather hail it as the creator and governor of the realm of matter.*[3]
>
> The universe is of the nature of a thought or sensation in a universal Mind. . . . To put the conclusion crudely—the stuff of the world is mind-stuff.[4]

Donald Hoffman

Cognitive psychologist Hoffman, of UC Irvine, thinks we perceive reality as a complex of symbols that hides a vastly more complex reality. Hoffman likens this process to a desktop computer interface. He gives the example of a novelist using a computer to write a book. They see an icon on their desktop that represents their novel, but the icon, and even the document, aren't the basic reality.

The underlying reality is a complex string of information, 1s and 0s that manifest as an electrical current through a circuit board, the complexity of which we never see. Hoffman is saying that the reality we perceive is a limitation of deeper consciousness. We see only its representations, not its underlying reality. This is similar to the Gnostics and Plato, who say that we exist by limited consciousness, perceiving only shadows of things in higher dimensions.

This also supports what we learned in chapter 16, about neurophysiology. We never truly see objects "out there." Instead, we see electrochemical representations in our brain. Hoffman says these limitations have a survival function. If writers had to manipulate binary code to write a novel, or hunter-gatherers had to be aware of the laws of physics to throw a spear, both would have gone extinct. A Darwinian slant there for sure, but he's saying higher consciousness must be limited in order to allow us the simplified shortcuts necessary to function in the grosser physical world.

Equate Hoffman's desktop-icon image with Gnostic archetypal images and the idea of experiencing reality through a state of limited consciousness, and you're talking a completely Gnostic view of consciousness and reality. Moreover, in another very Gnostic statement, Hoffman identifies the thing that supports the reality beneath everything:

> In like manner, we create an apple when we look, and destroy it when we look away. Something exists when we don't look, but it isn't an apple, and is probably nothing like an apple. . . . Well, then what is reality? If my dog is only a data structure indicating a furry creature that enjoys fetch and hates baths, then what lies beneath that representation?[5]

For Hoffman, the answer is consciousness.

David Bohm

Eminent physicist David Bohm, mentioned earlier, used the term "implicate order" to express that a more fundamental level of reality exists. He believed that we are seeing broken reflections (blips or frames) of this fundamentally ordered reality. Here, again, is the repetitive theme that we only perceive representations or shadows of a more basic existence. Bohm proposed the idea that consciousness was somehow a manifestation of this implicate order, and that attempting to understand consciousness purely by looking at matter in space was doomed to failure.

> There is a universal flux that cannot be defined explicitly but which can be known only implicitly, as indicated by the explicitly definable forms and shapes, some stable and some unstable, that can be abstracted from the universal flux. In this flow, mind and matter are not separate substances. Rather, they are different aspects of our whole and unbroken movement.[6]

What Bohm says here ties together two previous ideas we learned: that the material world is a blip or flux in the "smoothness" of the quantum field; and that the "flux" corresponds to the forms and shapes of (Gnostic) archetypal images coming from a higher-order Consciousness from which we are not separated but exist as parts of the whole.

John Wheeler

Wheeler was one of the most influential modern physicists. He is associated with the Participatory Anthropic Principle (PAP). Wheeler pioneered the idea of a "participatory" conscious universe in which all of us are co-creators.

> We are participators in bringing into being not only the near and here but the far away and long ago. . . . Useful as it is under ordinary circumstances to say that the world exists "out there" independent of us, that view can no longer be upheld.[7]

Interestingly, Wheeler's view of consciousness is restricted to human consciousness. I never found any opinions by him suggesting a transcendent ground consciousness, though perhaps somewhere he has alluded to that. From a Gnostic viewpoint, Wheeler is correctly describing how we created (or distorted) the universe from a psychic level, but he stopped short at recognizing that that, too, is a derived reality from a higher source.

Philip Goff and Tam Hunt

Philip Goff is a philosophy professor at Central European University in Budapest, Hungary. Tam Hunt is a psychologist at UC Santa Barbara. I mention them together because their ideas both reflect the views of *Panpsychism*. This theory holds that science can manipulate matter, but that it has no idea what the intrinsic nature of matter is—similar to what I pointed out earlier regarding energy.

They view consciousness as a fundamental feature of all physical matter, down to every single particle in existence. Panpsychic theorists, like Goff, insist that it is a materialist theory despite superficial appearances. It explains consciousness as an aggregation of lesser forms of consciousness building to more complex forms of consciousness, as we discussed in chapter 16.

> The view is that there is just matter—fields, particles—nothing supernatural or spiritual, but matter . . . in terms of its intrinsic nature, is constituted in forms of consciousness.[8]

Tam Hunt elaborates on this theme:

> The central thesis of our approach is this: the particular linkages that allow for large-scale consciousness—like those humans and other mammals enjoy—result from a shared resonance among many smaller constituents. The speed of the resonant waves [vibrations] that are present is the limiting factor that determines the size of each conscious entity in each moment. As a particular shared resonance expands to more and more constituents, the new conscious entity that results from this resonance and combination grows larger and more complex.[9]

This is in complete concurrence with what I said in chapter 16, about simple cellular consciousness aggregating into structures such as brains. It also reinforces the idea that frequencies are what distinguish levels of consciousness. From this point, the Gnostics (and this author) part company with Tam Hunt's view. Panpsychism, though somewhat moving beyond the dualism of pure materialism (mind vs. matter), stops short in proposing how consciousness arises in the first place. "It's just always been there" doesn't seem adequate for a theory that begs the question.

It's almost as if the theorists stopped short of identifying Universal Consciousness so as not to be subjected to the standard scientific criticisms about God, religion, metaphysics, and untestable voodoo. Though panpsychism moves a few notches away from dualism, it doesn't envision an all-encompassing Master Consciousness Unity, and so the problem of dualism remains. Several similar theories exist, but nearly all of them are limited by the materialist ceiling of traditional science that confines consciousness to the physical world.

SUMMARY

By now your head may be swimming with all the particles of information I've shot at you, like the electrons in a double slit experiment, so let's summarize this book's theory of consciousness based on what we've learned from the ancient texts, psychology, neurophysiology, and quantum physics.

- A transcendent Master Consciousness exists, which is the ground reality of all dimensions and phenomena, seen and unseen.
- This Master Consciousness projects or disperses other points of consciousness from Itself, but in limited degrees of awareness or vibrational frequency, to create states of individuality. This consciousness is present in all forms, such as unseen intelligent entities, energy waves, particles of matter, humans, and even inorganic objects.
- Dimensions are levels of consciousness. The downward flow of the Master Consciousness creates parallel dimensions containing

various forms of intelligence, as distinguished by their frequency or vibration—that is, the degree to which they retain the consciousness of the Source.

* These dimensions interpenetrate and affect each other. They contain different degrees of energy and matter, from the finer or more subtle to the denser or more solid.

* This downward flow of Master Consciousness is altered and modulated as it passes through the lower-level centers of awareness, such as by influences from the psychic dimension. The imagery of a mountain lake's pure water picking up increasing levels of contaminants (limitations) as it flows to its lowest point illustrates this condition.

* A kind of cosmic algorithm exists in a template of guiding concepts and images called archetypes that direct or influence our minds. The clarity and influence of these archetypes experience negative distortions in lower dimensions of consciousness, owing to the ignorance and sense of separation that increases the further removed a conscious entity is from its Source.

* Human beings are tripartite creatures following the paradigm of:

Spirit (Self) > Mind (Soul) > Matter (Body)

* The Self is our spark of higher universal consciousness, the mind is our psyche or unconscious, and the body is tied to the conscious ego. The lower levels have glimmers of the dimensions above them, so each dimension displays conflicting interplays of light (awareness) and shadow (ignorance).

* Humans are unconscious co-creators of physical life and the material universe. This is evidenced by how human consciousness collapses waves into particle form (remember the double slit). If, however, one achieves an integration of our fragmented natures, co-creation can become an increasingly conscious process.

* Energy, intelligence, information, and matter are all aspects of a singular thing—consciousness. That they appear to be distinct is part of the illusion that allows us to experience a world of seemingly three-dimensional objects moving in space and time

instead of a static, unchanging underlying reality of infinite oneness.

- Given all the points above, we have strong indications that we are living a simulated collective reality, largely by our own choice and making, that is being created by higher aspects of ourselves in order to experience materiality. We are like virtual simulation programmers lost in our own holographic simulations.

Let's close our discussion out with an observation by the esteemed Albert Einstein:

Physical concepts are free creations of the human mind, and are not, however it may seem, uniquely determined by the external world.[10]

27
Three Thousand Years Late
How Mysticism Predicted Quantum Physics

We have come to the point I promised you early on, where we directly compare the Gnostic Creation story with the findings of modern quantum physics. The results are most startling because, you could say, quantum physics was three millennia late in many of their findings.

It is important to understand a few things at this juncture. Mystical and spiritual visions largely come through symbolic images. No invisible narrator blurts out literal messages, saying, "Now we're going to explain the Big Bang to you!" The original Gnostics lacked the scientific vocabulary of modern technology. They had to describe quantum events mythically and allegorically. This means interpretation of their texts must be involved.

For this reason, some will inevitably argue that anyone can impute their biases or interpretations onto the Gnostic materials. First, let me say that the same criticism applies to science, which is why we have so many competing theories to interpret the same observable phenomenal data. Therefore, cross-referencing data from multidisciplinary information is a good way to see if we're on the right track. That's why the format of this book includes psychology, neurophysics, and quantum physics, as well as mystical wisdom to find common ground as an indication of truth.

The Gnostic Creation incidences that follow are described in bold, with bullet points giving their quantum correlation. The general elements of the Sophia story are mostly from the Gnostic text *Pistis Sophia*, from chapter 30 on. The Sophia story is also universally repeated in any other number of texts, so I'll mostly dispense with citations except for specific important points.

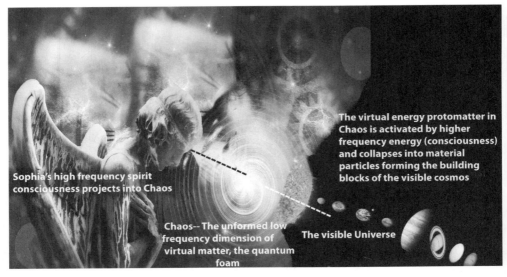

Figure 41. Scientific Diagram of the Gnostic Universal Creation

Sophia descends into chaos.

Sophia descended into chaos, where her high energy mixed with and activated the dormant elements of potential (proto-matter). This was the quantum field, and Sophia's energy signature caused ripples or blemishes in the formerly smooth energy field. These blemishes became the **virtual particles** permeating the universe as collapsed energy waves. They were like partitions, able to exist apart from the broken unity like one bubble forming off another bubble, as demonstrated in the double slit experiment in chapter 14. These "blips" form the quantum basis for the appearance of the material world.

Sophia breaks with the divine creative law and Source to pursue Creation on her own.

Sophia's action is a perfect example of **symmetry breaking**. Scientists believe that, just after the Big Bang, all the fundamental forces of nature were united into one force, but then broke apart one by one. This opened the door for the formation of matter and virtual particles, which could not form before then. Remember, Sophia's break with Source led to the formation of matter.

Sophia is surrounded and trapped by the elements in chaos, thus activating matter.

This is a stunning description of the "God Particle," or Higgs Boson, and how energy is converted to matter. Our universe is permeated by something called the Higgs Field, through which all energies pass. Upon contact with the Higgs Field, the Higgs Bosons (energy-carrying virtual particles) surround and attach themselves to the energy streams. They slow the frequency or vibration of the energies to the point where they take on the particle mass that forms matter.

Sophia, like all Aeons, is not so much a person as an intelligent, vast current of energy, or high-energy force field.

Upon contact with the proto-elements or virtual particles in unformed chaos, the proto-mass elements surrounded Sophia, drawing off or slowing her light energy like the Higgs Bosons. As her energetic frequency or vibration slows, again, as the Higgs Boson does, she is being converted to matter. She cries out to heaven in alarm:

> I became as heavy weighing matter . . .[1]
>
> I have become as lead. . . . When they had taken my light [energy] from me . . .[2]
>
> Save me out of the matter of this darkness that I may not be submerged therein . . .[3]
>
> Take not away thy light image from me. . . . For my time is vanished like a breath and I am become matter . . . [4]

The parallels here to the operation of the Higgs Boson couldn't be clearer. This part of the story accurately describes the process that created matter in nontechnical, mythical terms. The Gnostic seers defined the God Particle three thousand years before modern science.

The new matter resulting from Sophia's action was expelled, hurled away, and projected apart.

This is a description of the appearance of subtle matter during the Big Bang. The appearance of matter caused a disturbance in the unseen realms, most likely because spirit and matter differ in frequencies and cannot coexist in the same dimension. So:

> . . . matter came into being out of shadow, and was projected apart . . . expelled like an aborted fetus. . . . And it did not depart from chaos; rather, matter was in chaos, being in a part of it.[5]

The depiction of matter being pushed out and expelled to then reside in chaos demonstrates that a new (psychic) dimension had come into being that was filled with vast quantities of subtle matter, likely the same as dark matter.

A "watery" medium appeared in chaos, from which subtle or energetic matter was converted to visible matter.
This Gnostic mention of a watery medium sounds very much like the quantum foam coined by the great physicist, John Wheeler. Quantum foam is Wheeler's term for the energy field in which virtual particles exist as fluctuations, like bubbles in beer foam. From these fluctuations, visible matter arises like the blemishes, blips, fluctuations, and disturbances mentioned several times before, as part of quantum theory.

> For all of it [chaos] was limitless darkness and bottomless water . . . there appeared for the first time a ruler, out of the waters . . . he had come to have authority over matter . . . he saw nothing else, except for water and darkness . . . moving to and fro upon the waters . . . the ruler set apart the watery substance. . . . And from matter, the ruler made a footstool, and he called it "earth."[6]

In the passage above, the ruler is the Archon, the psychic entity we describe as symbolic of collective fallen soul-minds seeking individual experience in new dimensions of existence. The creation of the material world from this intermediate psychic level is what I've referred to as the contaminated lower levels of the pure higher streams responsible for material creation. You may note that the image of spirit moving over the water is also present in the book of Genesis.

The Archon's dimension did not obey the laws of the higher dimensions.

Scientists generally agree that the parallel dimensions are likely governed by different laws than our three-dimensional world. This describes the Archon's dimension:

> And he [the Archon-Demiurge] did not obey the place from which he came.[7]

• • •

I've just recounted the parallels between the Gnostic texts and quantum physics, and the way each has attempted to provide us clues about the Creation. You could seize on any given correlation and call it a coincidence or semantics, but when "coincidences" become so numerous, this should give any rational person pause. And that, you see, is what I hope the reader will do—think about the larger picture and the forces behind what life is about.

We've analyzed the Gnostic texts in light of both psychology and quantum physics. I've prepared figure 42 to summarize and correlate key concepts so you can see them conveniently, side-by-side, combining all three disciplines. (See page 186.)

GNOSTICISM	PSYCHOLOGY	QUANTUM PHYSICS
Sophia's agitation to know God	Human existential dilemma not knowing our origin or purpose	Energy seeking to fill a vacuum
Chaos	Unconscious mind	Quantum Foam
Sophia crosses boundary into Chaos	Individuation separation from Unity	Higher energy enters Higgs Field
Boundary	Mind Divisions—conscious, subconscious, unconscious	Parallel Dimensions divided by Membranes
Chaotic elements engulf Sophia, lower her energy	Limitation of consciousness, psychic mind arises	Matter created by Higgs Bosons
Matter expelled in Great Disturbance	Separation of Souls from the Unity	Big Bang
Birth of Archon-Demiurge	Rise of ego-personality limits higher awareness	Disruption of the Quantum Foam leading to materiality
Simulated world Patterned on higher dimensions	Archetypes as projection of underlying reality	Holographic Universe
Sophia's redemption from lower planes	Enlightenment, actualization of fragmented psyche	Theory that Consciousness will unify energy and matter and control underlying reality

Figure 42. A Summary Comparison of Gnosticism, Psychology, and Quantum Physics

28
Never Mind

Understanding Mind Levels

In earlier chapters, I spoke of ways we live in an illusory world. Another part of this illusion has to do with the way our minds process information.

We have three basic levels of human mind operation, not counting the higher self with which we're trying to communicate: the conscious, subconscious, and unconscious.

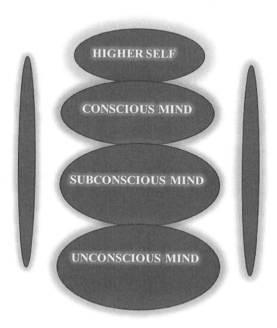

Figure 43. Levels of Mind

Let's use a computer analogy to help explain the workings of these three forms of consciousness.

The conscious mind is your everyday walking and talking mind. You are fully aware of the contents of this mind. It acts like the processing and keyboard input features of a computer. The computer keyboard inputs data into the processor and the results are displayed on the monitor screen. Your conscious mind works similarly. It absorbs information from internal or external stimuli and the results are processed and judged on the screen of your conscious awareness.

The conscious mind uses thought, verbalization (words), pictures, and writing to communicate. It is the directing, conceptualizing part of the mind. It is proactive, compared to the more passive subconscious and unconscious minds.

The subconscious mind is like an automatic recording device. It stores data accumulated over a lifetime, even data not immediately recognized by the conscious mind. The stored data of the subconscious is like the cache function in the RAM (Random Access Memory) of a computer, where programs and data frequently used and ready for quick retrieval are stored so they can easily be reached by the computer's processor.

So, information stored in the subconscious is not immediately needed, but readily retrievable—unlike information stored in the unconscious. Recent memories are stored in the subconscious for quick recall when needed, such as your address or the name of a city you traveled to last year. The subconscious also holds the needed programs you run every day, such as recurring thoughts, behavior patterns, habits, and feelings.

The subconscious is reactive, compared to the proactive conscious and unconscious minds. It responds to thoughts from the conscious mind and emotions from the unconscious. The subconscious is the faithful servant that carries out the intent of its bosses. It performs best when thoughts and emotions are in harmony.

If thoughts and deeper emotions conflict, the deeper emotions will win every time. That is why, when things go wrong people will say the result was not what they'd wanted, what they'd intended, or that for which they'd prayed. Conscious desires lose out to strong emotions.

Your unconscious mind is like the solid-state drive (SSD) in your computer, containing deeper background programs that need greater degrees of processing to access. Like a computer's SSD, the unconscious comes preloaded with certain programs from the collective unconscious, or perhaps from past lifetimes, acquiring other programs during one's current lifetime. Your unconscious mind is the long-term storage place for all your personal and collective memories and programs that have been installed since birth. The unconscious is the murkiest and most difficult to understand of all three mind levels. In part, this is because the contents of the unconscious express themselves automatically and are not subject to introspection or conscious recall.

Even though the contents of the unconscious mind remain below the level of conscious recall and awareness, the emotions, memories, and ideas grounded in the unconscious exert a tremendous influence over our lives and our behavior.

The unconscious mind is a paradox. It is the least understood, most mysterious, and most influential part of our psychic makeup. Paradoxically, it is simultaneously both our damnation and salvation. I know that sounds a bit religious and maybe a bit dramatic but, while it can be dramatic, there's nothing religious about it.

The unconscious is, rather, a psycho-spiritual bridge to higher dimensions. It is the repository for suppressed memories and traumas from our present lives, as well as karmic memories of past lives. It is the soul's source mind, where we hold the deep-seated beliefs and habits that control how we experience things. These crystallized beliefs will play out in our lives for better or worse, so it's wise to harness them for better life results. Communication with the unconscious is the function of meditation.

Information from the Wholeness, the Source that created everything and knows everything beyond space and time, is projected through the medium of the human mind, then onto the brain, to produce electromagnetic perceptions of matter. But now we understand why we do not experience pure vibrations of divine awareness—the knowledge of Source is fragmented and processed by the differing layers of recognition and perception in our multifaceted mind.

SURFING THE BRAINWAVES

Each of these levels of mind has corresponding brainwaves, such as Alpha, Beta, Theta, and Delta. Beta levels are higher frequency, usually associated with an aware, engaged conscious mind, like someone giving a public presentation. Alpha waves are lower frequency, usually associated with the subconscious. This is a resting state associated with meditation or a relaxed state of non-arousal. Theta levels are of even lower frequency, associated with the upper levels of the unconscious. This is the state of lucid dreaming or even daydreaming, not actually asleep but only partially aware or awake. It is associated with the REM (Rapid Eye Movement) state during sleep. Delta waves are indicative of deep sleep or dreaming, well below the conscious level.

Some forms of deep meditation can occur at the Theta level. Edgar Cayce, the "sleeping prophet," relayed amazing psychic information while sleeping, but had no conscious recollections of his messages when he awoke. He probably operated from the Theta or even Delta levels.

The Alpha level is best for meditation because we can be relaxed and receptive while remaining awake, enabling us to direct intentions and absorb insights. Here we cut out the "mind chatter" of our conscious engagement with material surroundings, and experience the nonmaterial inner dimensions from which we derived. *This is the level we use to affect change and assert a connection with our underlying creative power.*

OF THE RIGHT AND THE LEFT

Another process that contributes to our skewed view of reality is built into the very structure of our brain functions. Earlier we touched upon the left brain/right brain concept. This is essentially an energy orientation, a default vibration, if you will. Though the essence of reality is a unity, at the lower planes of existence it expresses as a duality, the union of polar opposites, the yin-yang force that permeates everything in the universe. The ancients generally viewed the characteristics in the chart in figure 44 as the male and female forces.

Left Brain	Right Brain
uses logic	uses feeling
detail oriented	"big picture" oriented
facts rule	imagination rules
words and language	symbols and images
present and past	present and future
math and science	philosophy and religion
can comprehend	can "get it" (i.e., meaning)
knowing	believes
acknowledges	appreciates
order/pattern perception	spatial perception
knows object name	knows object function
reality based	fantasy based
forms strategies	presents possibilities
practical	impetuous
safe	risk taking

Figure 44. Left and Right Brain Functions

According to this belief, the left brain is the logical brain that analyzes, while the right is the imaginative brain that intuits. The left brain *thinks,* the right brain *feels.* Though every person has traits of both attitudes within them, most of us default to one side or the other. Far fewer are truly balanced in the use of these faculties.

People too oriented to the left, for example, do not pay attention to or even value intuitive feelings. People too far to the right tend to disdain facts in favor of fantasy or abstractions, and often act illogically or impulsively with little forethought.

The imbalance of these attributes is another factor that contributes to the layer of illusion by not allowing us to penetrate the underlying unity of things. Most people today are stuck in left-brain thinking that focuses on logic and the primacy of material reality.

Jesus, in the Gnostic gospels, emphasized the importance of balancing ourselves. He said that we must make the male into the female and the female into the male to realize the kingdom of heaven, i.e., to achieve higher consciousness. To achieve the Gnostic goal of direct contact with divine information, the mystic seers had to hone the balance between their rational and intuitive natures. Meditation was, and remains, the prime vehicle to achieve this.

LIFE IN THE MATRIX

Well, let's see—our vision is constructed so that we never directly experience objects, the world appears solid but is not, our brains are split into two conflicting parts, and our minds into three levels of perception. You gotta ask yourself why on earth did whatever created us go to such extraordinary lengths to bury us under so many layers of illusion? What was it hiding? Even the ancients had to tackle this question. Plato illustrated it with his allegory of the caves (see figure 45).

Absolute truth or true reality is depicted by the sun outside the cave. Inside the cave, humans are chained to a wall to face one direction only. A false light, depicted by the fire, casts the shadow of a potted plant on the cave wall. The chained people can only see that shadow, so they take it to be their reality. In shorthand, we're imprisoned by false sensory perceptions.

This ancient Gnostic/Platonic image perfectly reflects the illusory states and the fact that we never directly experience objective reality. The key question is who or what is that fire in the allegory that manufactures the false reality we perceive? Well, if you've been paying close attention, don't blame God. The answer is you and me. We chose this experience. The fire is the Archon, the collective psychic mind force that we as fallen spirits used to dream ourselves into the illusion of a material world.

Think back to earlier chapters: Only One Substance, One Energy, and One Being exist. The only way this Being can create identities outside Itself is to limit Itself, forget Itself, and let the individual mind-children It projected think they are separate from Itself. To do this, a different mind reference, an ego-mind if you will, is necessary for self-aware individuality, both for Source and Its projections. If Source and Its projections were all of one unlimited mind, no individuality would be possible.

That, my friends, is why illusion is necessary. Life is a descending

Figure 45. Plato's Allegory of the Cave

series of deeper illusions, ranging from worlds unseen down to the basement we call material existence. We chose, at a psychic level, to pretend we were material creatures with physical senses that detect material objects, but the truth is that we created a quantum-electric 3D light show where we, the actors, began to think we are the characters we are playing. Like the allegory of the caves, we are shadow copies or "likenesses" of a higher reality. See this passage from the Gnostic gospels:

> And having created . . . everything, he [the Archon/the collective us] organized according to the model of the first Aeons [heaven] which had come into being, so that he might create them like the indestructible ones. Not because he had seen the indestructible ones, but the power . . . produced in him the likeness of the cosmos.[1]

As one Gnostic observer on the internet put it:

> It will not be a coincidence that [the Archon] came to be, but a preparation for the exodus of souls from God's light world. As the Gnostics

taught, we all existed from the very beginning in that light world. However, some of us wanted to go out of it to have experiences that are not possible to have in the light world—and to fully experience our free will (also in ways that are impossible in the light world). For this reason, God created various angelic levels and then a region of darkness for those who were to fall still further down. To create space for these levels below the divine light world, God contracted himself.[2]

CHOICES IN THE GARDEN
OF GOOD AND EVIL

Feeling trapped in a shadow world of suffering and illusion can create feelings of futility and depression. Truthfully, if we give up and remain in this illusion, our separation from Source becomes greater, and evil eventually sets in. Evil is extreme selfishness that worsens as we weaken the connection between ourselves and our Source. We become distanced from the concept of unity between ourselves and others, so we think less about harming others. We forget that by harming others we harm ourselves; it's only us and our desires that count.

But another path exists, and it is beautiful, magnificent, and sublime. We've been wrongly taught that humans are born sinners who have offended God. We are lowly creatures stumbling around an earthly insane asylum trying to get back into God's good graces—born in sin and playing catch-up all our lives. It's not a very hopeful or pretty picture, so don't tread that path because another way is open—the glorious way we were meant to follow, the end game for all the suffering we have experienced.

You are critically important in the grand scheme of Creation. Without human consciousness receiving, transmitting, and observing light energy, the universe would not even exist. This "trapping" of our consciousness in a lower plane has made possible an experience unfathomable in higher realms—the experience of immaterial spirits having a material experience.

We're not just finite physical forms—we contain the spark of the divine spirit within us. All the previous information in this book demonstrates that humans are the bridge between heaven and earth, creatures of both worlds. This is a high state, not a lowly one, and it places humanity above the static consciousness of higher beings, in terms of breadth of experience. This is likely the origin of the legend of Satan and the angels' jealousy of mankind.

I passionately believe, and I try again to convey to you in humble words, that *the purpose of humanity is to spiritualize the material and bring the experience of material back to spirit so that all the planes of existence align in harmony, and all separation dissolves for the illusion it is.*

I was thrilled to find my observation supported by a passage from the Gnostic gospels:

> Blessed is the man who has known these things. He has brought heaven down, he has lifted the earth and has sent it to heaven.[3]

This is the exalted destiny of mankind, what the Bible called the New Jerusalem and the Kingdom of Heaven on Earth. We will know ourselves to be ourselves, yet know ourselves to be our Source, simultaneously. We will be able to flow back and forth between spirit and body, as the most ancient legends say early spirits on Earth were able to do before succumbing to their own created illusions. In scientific terms, we will evolve (or revert?) to a state where bodies are not necessary to experiencing material consciousness, and our consciousness can either transcend or experience bodily existence at will, participating in different dimensions inside and outside of space-time.

DON'T WORRY

We have been needlessly weighed down, as a Judeo-Christian culture, by the concept of guilt arising from sinning in the eyes of God. This is more dualistic rubbish. No need for forgiveness actually exists because evil and sin are at their root dualistic illusions happening in a dream projection. The only sin is the mind-error of thinking we are something other than Source, the ignorance of believing there is an "us" and a "Him."

This isn't an excuse to go out and do bad things, since action contrary to unity and harmony perpetuates our dream illusions. Our true condition is a unity, so the only sins are against ourselves. If we do harmful, unloving things, it keeps us in a longer state of suffering and misery in this lifetime, or others (karma). We see others moving ahead of us in happiness and well-being and, on a spiritual scale, our classmates graduate while we stay stuck in a dream-world detention of our own making.

I believe the grand vision of humanity's potential is the logical conclusion that follows the mystical, psychological, and scientific information we've covered. If you don't agree, don't worry. It won't affect your ability to experience great benefits from the next sections, which get into the nuts and bolts of how to make all this information work for us.

29

"What About Me?"

Conditioning Yourself for Consciousness

By this time, you've read through some pretty heady stuff. All the knowledge and ideas presented here are, hopefully, a helpful guide to tackling life's big questions—who are we, how did we get here, and what is reality?

The real purpose of this book is to show you that higher information lies dormant within you, and how you can contact it. You can develop the ability to directly contact the divine—this was the message of the ancient, mystic Judeo-Christians.

We need to discuss some issues and principles in preliminary preparation, to set the stage for having extraordinary experiences. Having a grounding in these issues is similar to all the background theory you're taught in school that prepares you to achieve specialization. In this case, that specialization is *your higher consciousness*. That means moving beyond the intellectual knowledge you derived from this book to direct experiential knowledge of higher consciousness.

FAITH VS. KNOWLEDGE

Is faith a good thing, like the religions teach? Short answer—*faith is good but knowledge is better*. Faith is a good stepping-stone to higher awareness, and that's how it was viewed by the original Christians. But the problem with faith is that you are one step removed from experiencing the thing you're asked to have faith in. Faith is relying on the word of an institution or individual to tell you about the way something is or was.

Faith can also be misleading. Dispensers of faith can be wrong, they can misinterpret, or they can be downright devious and distort things. It wasn't long into early Christian history before certain people found power and authority over others by becoming arbiters and guardians of the faith. Faith can be manipulated. After all, how can you disprove it?

Here's the problem from a spiritual standpoint—*prophets speak the heart of God, their followers start religions.* What separates them is experiential knowledge.

The followers do not have the same revelatory experience as the prophet. Therefore, they start interpreting. In interpretation lies the potential for the abuses of our human frailties to take hold. Like political correctness today, people seek to make their interpretation the undisputed universal standard for all.

Whose word would you rather take, the prophet's or the follower's? Better still, would you not want to become your own prophet? Don't get me wrong—there's tremendous value in scriptures and traditions, but there is also misunderstanding and misinterpretation, often deliberate, to promote one agenda or another. If you want to know the truth, Get it from the Source's mouth (pun intended).

Knowledge comes in two forms, intellectual and experiential. Intellectual knowledge is learned knowledge, or the mere accumulation of information. Experiential knowledge derives from actual intercourse and familiarity with the subject matter or source of information. The difference between the two is like being told that fire burns versus sticking your hand in the fire. This dispenses with having to take someone else's word on the subject because you've *directly known the subject* in some form.

We can identify a further subset of experiential knowledge— mystical knowledge. This is experiential knowledge derived not from ordinary but from *extraordinary* experiences, such as the kind the ancient seers spoke of in their altered states.

Like the findings of quantum physics, this type of knowledge is often counterintuitive to normal perceptions, but it directly proves something profound to the person experiencing it, about the illusory nature of our sensory world and the true nature of underlying reality.

WHAT'S THE PROBLEM?

You may say I'm no psychic; I've never had any extraordinary experiences. Why should I believe in them? Several reasons can account for that attitude.

- You aren't seeking extraordinary experiences.
- You don't believe in extraordinary experiences.
- You don't have a firm goal or reason to have extraordinary experiences.
- You might be afraid of extraordinary experiences, given negative connotations (evil, voodoo, occult).
- You're too focused on the ordinary routine to have extraordinary experiences.
- You were conditioned to believe the world is just what you can sense and nothing else exists, that there are no extraordinary experiences to be had.

This last point is widespread in our culture, given our historical trend to linear, logical, left-brained analytical approaches to life. According to one source, Albert Einstein was aware of this:

Albert Einstein called the intuitive or metaphoric mind a sacred gift. He added that the rational mind was a faithful servant. It is paradoxical that in the context of modern life we have begun to worship the servant and defile the divine.[1]

SPEAKING FROM EXPERIENCE

Probably 90 percent of what I'm going to tell you from this point on comes from personal experience. I started having these experiences before understanding the principles or causes behind them. I had to study and grope my way toward these principles, and now I'll relate them to you as best I can.

Motivation

We all need a starting point and, to embark on any quest, we need motivation. This can take numerous forms. Sometimes we hit rock bottom in life and this causes us to look in other directions to make sense of our suffering. Some of that was present in my case, but more of it was a desire to help people. I felt that, in doing this, I would be helping myself in some undefined way while being of some practical benefit to the larger world. I must tell you honestly, though, more often than not it takes an internal or external crisis to motivate people to start a real inward journey. I suggest you read books, attend lectures, or talk with experienced people as a gentler means of finding your path.

Desire and Direction

Desire works hand-in-hand with motivation. The difference as I see it is that desire is a gut emotional feeling whereas motivation is about goal-setting intentions deriving from the intellect. I believe feelings are more powerful than ideas. Pursuing ideas sets useful goals, but ideas can become an abstraction. Pursuing deeply felt feelings has more legs for the marathon.

That said, both desire and motivated direction are needed to launch your personal quest for the extraordinary, and launching shall be our metaphor. Picture our starting point as a rocket ready to take off. Rockets need two main systems in order to function, the fuel and guidance systems. Desire, a right-brained quality, is the fuel that burns and gives the energy of propulsion to lift off. Direction, a left-brained quality, is like a logical set of algorithms in a guidance system telling us where we want to go and setting goals we want to achieve.

Both are necessary to get us to the place we want to be. Without the fuel of desire, we'd never get off the ground. Without the guidance of direction, we'd lift off but spin around in circles.

Intention

All manner of intentions can lie behind our motivations. Entire New Age industries developed, starting in the 1970s, using spiritual principles of co-creation to achieve personal gain—money, cars, relationships—

not that anything is wrong with that, but results were often lacking. The reason is intention. I have no rational scientific basis for what I'm about to tell you, but I do have something that trumps that—personal experience.

Good intentions work better than selfish ones when it comes to manifesting results in your life. The guy who wants a new car to pick up women more easily will probably wait longer than the guy who wants a car to get to his healthcare job to treat his needy patients. I know it sounds corny, but the ideals of love, kindness, service, goodwill, and unselfishness go a very long way when it comes to opening higher channels of energy and consciousness.

Alignment

Alignment cannot happen when your conscious intentions or beliefs conflict with the deeper emotions of your unconscious or subconscious minds. For example, your intention may be to develop a relationship with a particular person, but your deeper feelings warn you off. Your desire can only be realized when your feelings are in alignment with your conscious intent. Go with your gut in such cases.

Humility

When extraordinary phenomena first begin to manifest in your life, it's a pretty trippy experience. Our ever-present egos can enter the picture quite easily. Bending spoons in front of your friends, giving them accurate information by touching a personal object of theirs, or drawing a picture they hold hidden from you with unquestionable accuracy is sure to elicit the oohs and ahhs that tell you how special they think you are. Sometimes these demonstrations have their place, but often they are just ego-driven parlor tricks, a pitfall to those who have developed some degree of awareness. If you possess the proper attitude of humility, you will use your gifts in ways that are appropriate and beneficial.

Gratitude

Close on the heels of humility comes gratitude, a genuine feeling of thankfulness that you have connected with a greater power that is guiding you, sustaining your faith, and letting you know that whatever you're experiencing in this life is not the be-all and end-all of existence.

Imagination

A great way to trigger extraordinary experiences is through imagination. In the proper receptive state, which we'll discuss in the next chapter, we're putting the mind to work, galvanizing it to pull all the circumstances together to achieve our goals. In this state, the subconscious mind is more programmable. What is imagined with feeling behind it is what becomes real. This allows us to hold a belief in our being and take necessary actions to change circumstances despite external appearances. Imagination then pulls the process along by opening up alternatives, rather than pushing from behind with wishful thinking. Imagination here works closely with *expectation*.

Expectation

Very important. Here is where we exercise our faith and imagination. Conventional prayer, which petitions God for favors, doesn't work because the premise is that God is separate from ourselves—a faulty premise. *We are part of the Master Consciousness,* and our desire is already fulfilled, just waiting to manifest. Visualizing the end result of your desire and expecting it to manifest galvanizes consciousness to flow toward it. If you can generate the feeling and expectation of your goal being achieved, the result is more likely to manifest.

Letting Go

You can set a conscious goal or intention before meditation and let your mind move in that direction, but at some point, you just have to let go and not overly dwell on your desire. If you overly concentrate on something, you let your conscious mind enter the picture, and that will interfere with the subconscious "autopilot" through which the higher sources operate. Reinforce your vision during meditation, then try to forget about it. Ever heard "let go and let God?" It's good advice.

When I used to do spoon and metal bending, we would start by consciously concentrating on the metal heating up and getting pliable, but it would only bend when we let go and shifted our minds onto something else. It was like a distraction that took our conscious minds out of the picture. Worrying or forcing something along brings the conscious mind to the fore and makes your non-ordinary mind take a back seat. You have to get into a mindset of detachment over the thing you desire, in order to attain your goals. The higher power knows what you seek, you don't need to keep telling it.

Visualization

A great way to tie all these things together during meditation is with visualization. You know how a great movie can involve all our senses and emotions? Become your own screenwriter. Construct an internal script and put images to it that exemplify the results you desire. Whenever you meditate, run this movie through your mind, then let go and relax.

Be Mindful

One thing to keep in mind is that you can influence results but you cannot dictate the form of the result. That car you've been visualizing may not come in the form of a Ferrari but as an unexpected deal on a car you can afford. That is because, to paraphrase the Rolling Stones, the Universe is aware of what you need more than what you want. Results often come in flashes of opportunity on which you need to act as opposed to waiting for results to fall in your lap, so be alert and mindful.

The points above are useful for positioning yourself with attitudes that can help you attain non-ordinary states of consciousness. Now, let's move on to developing your own personal platform for communicating with higher information.

30

Brass Tacks

Techniques of Meditation

The two previous chapters were like boot camp, where you got the tools, training, and conditioning to get out and perform in the field. You learned about attitudes to help put you in the optimum frame of mind to start your quest. You learned about the frequencies of mind and the state of physical relaxation most conducive to a receptive condition. You learned about specific obstacles you might face that create illusions. Most importantly, you should now have a vision of the potential rewards and purpose behind absorbing this material and starting your own contemplative practice.

No instant steps exist for awakening higher consciousness. You won't have extraordinary experiences just because you read this book. You didn't learn to ride a bike because your father told you how to do it—you had to go out and give it a few tries to get your balance. But learning anything new becomes much easier when you understand the principles and dynamics that govern your task. All the information in this book is like the helpful tips Dad might have given you before you mounted the bike.

Embarking on a journey requires some background knowledge and vision before you start down the road to discovery. Any successful adventure is realized through preparation, and that's what we've been doing up to now, so let's do a quick recap:

Everything in creation derives from a single Master Consciousness.

- This Consciousness operates through an energy field upon which it projects a variety of virtual experiences in differing dimensions or frequencies of existence.

- Each of us has a higher consciousness vibrating in a higher dimensional frequency, known as our Higher Selves.
- Our Higher Selves have used the energy field to project the experience of being in a physical, human form that has forgotten its true nature.
- Operating out of lower frequency ego-mind, we identify with our body, believing we are separate beings though we are still part of the chain of Master Consciousness.
- As part of this Intelligence, we can affect the energy field, manipulate matter, and organize reality albeit at an unconscious, low-frequency level of awareness.
- Through self-reflective techniques such as meditation, we can rise to the higher mind levels from which we originated, receive guiding information from extraordinary sources, and exercise more conscious co-creative control over our lives.

MEDITATION

Everyone has heard of meditation but few practice it—and those who do experience varying results.

Not all meditation techniques are the same. Some present a rather passive go-with-the-flow technique. Active meditation, by contrast, begins with a set intention or goal toward which you would work. Meditation is a bit unpredictable too. But the mind is all-seeing. Even if you choose a passive approach and don't set an intention, your mind knows what's bugging you, so answers might just pop into your life anyway.

I don't promote any particular meditation method, but I will relate what works for me. You can take or leave this information depending on how it resonates with you, but do develop a practice to realize the benefits of all the theory you've just learned.

BEGINNING

All meditation begins with some form of induction, usually accompanied by deep, rhythmic breathing. I often count backward, from 10 to 1. Try visualizing an elevator descending from a higher floor to the basement, mentally counting down the floors, and repeating the phrase "going deeper and deeper" with each level.

Meditation and the information that comes from it is a subtle thing. Unlike Hollywood portrayals of altered states, lights don't flash and bells don't ring. You won't really feel much different, aside from being relaxed. I borrowed a technique from self-hypnosis, which I used early on to ease my doubting logical mind.

After the induction, visualize a huge hot air balloon hovering above you with a thick rope attached to your wrist. See the balloon floating up, imagining it pulling your wrist and arm along—involuntary muscles may actually lift your arm high into the air without conscious effort. Once this happens, you can be fairly sure you're in the right meditative/brain wave level. Then, just mentally cut the rope and let your arm lower or flop back down.

CONTEMPLATING

Sit in silence for a few minutes, following your breath and keeping your mind as blank as possible. If you just want to relax, stay in this state. If your set intention is to receive information, visualize a bright stream of light from your head, running off into space, cutting through imaginary dimensional circles like a knife acting as a receptive conduit. I always surround myself in protective light, mentally repeating that I want only the highest and best information. I don't want some lower-level intelligence cutting into the broadcast. If I want to accomplish something, I run a movie of it through my head, visualizing the outcome I am seeking. This is where imagination and expectation come in.

LISTENING/RECEIVING

Meditation is not a process that provides literal answers. Sometimes your mind will verbally articulate information, but the language of the mind is mostly symbols, insights, and feelings. A warning here—flashes of insight come rapidly and can be dismissed by your logical mind as soon as they arise. This is your conscious mind intruding on the subconscious.

Your materialistically conditioned conscious mind feels threatened by any information that contradicts its illusions. Like your high school math teacher wagging a finger at you for imaginative daydreaming in class, it thinks your insights are worthless, impossible, or even dangerous, so it shoots them down very quickly.

Achieving results from meditation is more often a process, not an instant gratification. The power behind meditation doesn't lay full-blown solutions on your doorstep like a Christmas package. But it does pull elements together that cause a shift in the patterns you want to change, often manifesting as a series of steps or incidents that ultimately change your condition and get you to your goal. Meditation starts to move barriers behind the scenes.

We live in the "me age" of instant gratification, but that's the part of the ego mentality that caused our problems in the first place. You have to train yourself to be mindful and alert, to catch fleeting thoughts and insights. Look for patterns and synchronicities, and pay attention to shifts in attitude about things that were formerly firm convictions or behaviors on your part.

Miracles, my friends, begin as subtle shifts in attitude that lead to major changes in circumstances. We've heard of drug addicts or criminals who became ministers in service to others, or poor kids who rose from the slums to land a job on Wall Street—these transformations resulted from small shifts of mind, belief, and attitude, and the rewards came incrementally. Meditation is a tool that can align universal creative power with our desires and abilities, but it works on its own schedule and in its own ways. The time it takes to manifest things is related to *you* aligning with *it*.

I've spoken about possible results of meditation in generalities because everyone will experience benefits and miracles in different ways. But I do want to relate examples of my own experiences, to demonstrate that results can be quite real, helpful, and concrete.

Boosting Creativity

A while back, I wrote a novel called *Pope Annalisa*. It was challenging in several respects, not the least of which was striving to place a miraculous figure in the most realistic and accurate of real-world circumstances. So, much of my meditative focus was related to issues about producing the best book I could write. Scenes relating to the geo-political background necessary to the novel began to pour through me like automatic writing.

A magazine later did an article on the book, noting a substantial number of predictions that came to pass in subsequent years. A second Gulf war, Iran on a path to developing nuclear weapons, and the unexpected rise of a reforming non-European pope from a third-world country were a few examples of prescient information that funneled through my consciousness. The book was praised for its realistic and exciting backstory. I almost felt like I cheated, like someone else wrote large parts of the book—so strong was my sense of channeling the information.

Africa

Another striking incident was described in *Chicken Soup for the Soul: Dreams and the Unexplainable*. I was in Africa with my father, in a country that had some fairly lawless areas. Air travel was very restricted and we had an important meeting about an eight-hour car ride to the north. Our only alternative was private transport, which was usually some kind of small van that could seat around six people.

When it came time to get in the van, I recoiled and told my father something was wrong. I refused to get in the van. He was upset and we had some words because the meeting was important, but he had witnessed some of my unusual displays before so he reluctantly relented. We subsequently learned that the van had been attacked by brigands on some jungle road and a passenger had been killed.

Living in Africa was difficult back then. I struggled with feelings of fear and isolation. One day, on the streets of the port city of Lagos, an African man stopped me and just started talking to me as if he had known me all my life. He described what I was going through spiritually, including past and recent events, and he was encouraging me. It was like talking to a comforting friend I had known for years, and was just what I needed at the time. Later on, I learned he was a "bush doctor"—a kind of African shaman. I called him a spirit guide in the flesh.

Europe

A less dramatic incident happened one time when I had to fly back to Europe with a stack of important papers I needed to deliver to a bank. I disembarked but, just a few feet from the gangplank, I had a flash that three pages, 21–23, were missing. I waited until everyone was off, and they let me back on the plane. Sure enough, we found the exact pages underneath the seats.

Rome

One time, under supervision using certain breathing techniques, I had a vivid past-life experience as an Italian doctor in the 1930s who'd run afoul of the fascists for supporting Jews. I saw the street and the building I lived in, and even knew the street name and number. The street I saw was Via Virgilio Orsini. I knew it was on the Vatican side of the Tiber, near the river, because in my vision I could see the dome of St. Peter's a few blocks away.

I ran out to the bookstore and bought a detailed street map of Rome. Sure enough, I found the street situated exactly between the Tiber River and St. Peter's. The only inaccuracy was that the street was called Via Virginio Orsini, not Virgilio Orsini. I'd been off by one letter.

A couple of years later, my wife and I took a boat cruise that stopped in Rome. The first thing I did, on a day of pouring rain, was drag my wife to the location. Now, I even had pictured the building's address in my earlier vision, and a couple of other details that made it identifiable. Most of the buildings on the street were three stories or higher, and their front entrances were flush to the sidewalk. The building in my vision was two stories, and it was the only one with a small walkup, and that's exactly what we found.

We waited outside, under an umbrella in the driving rainstorm, hoping a resident might come out of the building so we could ask if they had any memory of a doctor who'd once lived there. No one came out in the foul weather. The building was gated, so we couldn't get to the front door to knock. We didn't speak Italian, so even if we had access, it would have made for a weird encounter. However, before we left, we saw a plaque on the gate with several names, all of them doctors. It was a medical building.

Coincidence, imagination, or pure baloney—you be the judge. But when the coincidences start piling up, you have to take notice.

I could relate numerous stories, but the gist is, uncommon experiences became so commonplace for me that the idea of other dimensions of existence ceased to be speculative and became a firm reality. In other words, as the Gnostics would say, I *knew* they existed because I had experienced them. I had passed from faith and imagination into direct experience.

The purpose of this book is to help you open up new dimensions of mind, in order to have extraordinary experiences. I've provided multidisciplinary information to give you the mystical and scientific background to make extraordinary experiences theoretically possible, demonstrated techniques to put them into practice, and examples of results that can occur.

I believe "unexplainable" phenomena become more intelligible and real to us if we have scientific and spiritual background to define the arena in which you can experiment with the game of life, meaning touching dimensions of information both seen and unseen. Your miracle stories will be different from mine because we all have a different focus and set of goals to achieve, but we can work with similar tools.

The ways miracles manifest are numerous and depend on your personal focus, goals, and orientation. But I can tell you that diligence with a meditation practice pays off, and your creativity and problem-solving capacities will surely multiply to improve your life.

Creativity and problem solving are connected. Both qualities stem from intuitive impulses and subconscious connections with ideas and information that seem to come out of nowhere, and maybe that's an accurate description because, as I said earlier, nowhere is exactly where we came from.

31

Is It Real?

Further Evidence

The human mind is plagued with doubts about things it can't see or information it is repeatedly told cannot be "real."

Is there really a Source of which we are a part? Are we really interconnected points of light-energy consciousness? Is the world around us really a virtual reality? The only way to truly experience these things as reality is from within the realm of your own mind, and it is up to you to study, meditate, and contemplate. This is important because many things exist that the common senses don't perceive, but that the intuitive mind can still detect and comprehend.

However, the intuitive right-brained mind is not the only level of mind we possess. We have to reach the rational left-brained mind as well. The intuitive mind is the spiritual mind, the rational mind is the scientific mind, and what I've done in this book is to reinforce their synergies by showing that they're indicating the same information from differing angles.

As I write this, scientists at the Brookhaven Labs in Long Island, N.Y., have just created matter from light energy, proving the counterintuitive fact that material objects are merely frequencies of light wave energy, just as Einstein predicted with $E=mc^2$.

Science has also shown the existence of infinity—a concept always associated with God or a Master Consciousness. Let me repeat—*science has shown the existence of infinity* not as an abstraction, but as a reality with mathematical proof, and this proof was described—you guessed it—by the Gnostic masters, thousands of years ago.

Here we have another case where the ancient Gnostic gospels

described phenomena that quantum science has only begun to uncover in the modern age: a scientific breakthrough concerning the phenomenon of fractals, which has been dubbed *God's Thumbprint*. It reads like a direct retelling of the Gnostic Creation story and the pattern through which the Creation operates.

FRACTALS

Fractals are repetitive patterns or structures found in nature that are similar from the largest to the smallest scales. They have been observed to exist for some time. Look at the image in figure 46, of a Romanesco cauliflower—see how it's composed of repeating structural patterns.

Figure 46. Fractal Patterns in Nature

Fractal patterns are present in all of nature, including plants, vegetables, animals, geographic features, and even in the organic structures of human bodies. In 1980, a mathematician named Benoit Mandelbrot, when studying fractals, came up with a revolutionary formula ($Z=Z^2+C$). Scientists call such formulas "elegant," meaning they are simple but describe profound things, like how Einstein's $E=mc^2$ described the relationship between matter and energy.

Mandelbrot's equation describes how fractals operate by *iterations*. In science, iteration means a repetition of a mathematical series where each version, or iteration, produced is applied back to a constant equation to produce a new version. For example, we can say this book underwent several iterations or versions from the original draft, each an application of the constant (in this case the author) who alters the previous iterations. The result is a number of manuscripts similar to but differing from one another.

Inspired by the repetition of fractals, Mandelbrot's formula creates an infinite series of similar iterations, each coming from the iteration that preceded it. The formula not only describes infinity through mathematical fractal iterations but also reduces it to a geometric pattern called the Mandelbrot set, as pictured in figure 47. If you zoomed in on the image, you would see infinite replications of this same image at differing scales. Like a hologram, each small part contains the image of the whole.

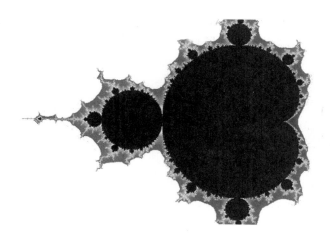

Figure 47. The Mandelbrot Set

GLIMPSING INFINITY, GOD'S THUMBPRINT

The Mandelbrot set of equations, embodied in figure 47, can be magnified to scales larger than the entire universe. New patterns will always emerge from the previous pattern, similar but subtly different from the one preceding it. This sequence goes on *forever*.

Though it can't be touched, the Mandelbrot set is a real thing, a real phenomenon. Anyone studying it can describe the same thing, so it is a universal experience. It's a geometrical shape and equation that embodies how the universe works. It may very well reveal the secrets of the Creation, which becomes more apparent when we introduce another variable—the

Gnostic Creation story. Before elaborating on this statement, I have to explain and simplify the elements of the Mandelbrot equation.

$$Z=Z^2+C$$

The letters represent coordinates, but let's imagine them as things, say a basket containing something. The letter C is a constant, so let's say it's the original, unchanging Master Basket. The first of C's mass-produced baskets is the letter Z.

The equation is saying that if you plug in a value or number for the first Z basket, multiply it by itself, and then add it to the constant of C, you get a new version or iteration of the first Z basket, though the contents will be slightly different. If you then take that new basket and repeat the process, you'll get a third version of the Z basket, again with slightly different contents. The process can be repeated indefinitely. Each time, it will produce a new version that will be nearly, *but not exactly,* the same as the previous version. Similarity, as opposed to exactness, is important here.

The Mandelbrot equation describes coordinates. It's a process of adding and multiplying, where one iteration must go back, multiply it by itself, and add that to a constant or source number to produce a new pattern or equation. It's an infinite loop process where the result of one operation becomes the input of the next.

For clarity, let's take a second example. Say the letter C represents a machine that is stamping out plastic molds and Z represents the batches of raw materials. Now the process introduces a new dye into each mold, which is subtly different from the dye used for the previous mold. The result would be a series of molds or versions very similar but with slight differences.

The Mandelbrot set is a geometrical shape and equation that embodies how the universe operates.

TWIN THUMBPRINTS

The Gnostic process of Creation is most explicitly described in *The Apocryphon of John,* or the *Secret Book of the Disciple John.* God, Spirit, the Source (call it any name you like) generates or emanates a series of

Aeons, independent spiritual entities similar to Christian angels. In actuality, they are currents of intelligent conscious energy that represent aspects of Source.

These aspects have names like Truth, Love, or Wisdom, indicating they are archetypal images that permeate creation and the human mind, as described by both the ancient Greeks and modern thinkers such as Carl Jung. You can also think of the Aeons as dimensions of existence possessing different frequencies that are variations of the original.

The Mandelbrot equation is an exact description of how the Aeons were generated. Each Aeon is like a variation of Z in the equation, while the Source is the constant C. Each "Z Aeon" has to return to the Source to gain Its "consent," or agreement as *The Apocryphon of John* describes it, whereupon it generates another Aeon. Remember how the Mandelbrot set is a loop process, where the output of one operation becomes the input of another?

Each Aeon loops back to gain the consent of the "the Source C" constant, whereupon it generates from itself a new Aeon. This process continues in the exact iteration described by the Mandelbrot equation. Each new Aeon is similar but not exactly the same as the one preceding it. Each Aeon contains less consciousness than its parent in a matter of increasing degrees because they are copies of copies. They degrade in fidelity and clarity, having ever-lowering energetic frequencies.

This is the reason why I emphasize the term *similarity*. The Gnostic texts reflect *similarities* among the Aeons, not exact reproductions. This is faithful to the Mandelbrot equation to the most minute detail.

WHO'S IN, WHO'S OUT?

Another remarkable parallel exists between the Gnostic gospels and the Mandelbrot geometric pattern. Depending on the number input to the equation, we get two different results. The number sequences either get smaller, in which case they fall within the organized pattern of the equation, or they get larger and fly off outside the Mandelbrot set into random chance, chaos, or complete freedom—meaning the choice to stay in harmony with our Source or to exercise our own intentions.

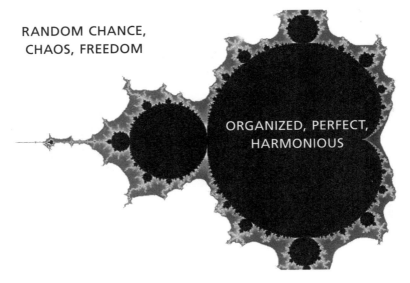

RANDOM CHANCE,
CHAOS, FREEDOM

ORGANIZED, PERFECT,
HARMONIOUS

*Figure 48. Mandelbrot Set Reflecting
Pattern of Creation*

Recalling the central Gnostic Creation myth, Sophia (Z) broke the divine law or pattern by generating without the consent of the Source (C). In other words, she ignored the expected return to the constant or original number, and violated the Mandelbrot equation. She substituted the prime number and entered her own funky wild card number into the divine formula. As a result, Sophia broke off into chaos. She generated an imperfect Aeon, a distortion that flew off the charts and resulted in the material creation many Gnostics and early Christians considered to be the outer darkness. This is depicted in figure 48.

Inside the Mandelbrot set is a perfect, harmonious, infinite whole of self-similar patterns, which sounds very close to the Gnostic descriptions of the Pleroma (Fullness or heaven). Outside is the world of random chance, chaos, and freedom, which sounds a lot like our world. Sophia's story is about breaking away from the ordered pattern of heaven in harmony with God's will to exercise her own desires. It is the story of exercising free will but also of experiencing all the suffering that comes along with our ignorance and misuse of that gift.

Infinity is like a loop upon which we ride in the light stream of creation, having different experiences in different states of creation. You could say Sophia's "sin," in Mandelbrot terms, was falling out of sync, and our purpose is to get back in the loop, as the saying goes. As to that task, the Gnostics and other mystical traditions promise that all things eventually return to their source. We can realign our numbers by realigning and awakening our consciousness, which, however degraded, still contains the spark of the force that is behind all creation.

This is why the Bible said we are made in the image of God, not the same but with a similar pattern of consciousness that we can purify and make more conscious, more godlike. The ability to self-reflect on our experience is what distinguishes us from lower forms of consciousness.

Here is a summary:

• Material creation is a play of light-energy frequencies.
• These frequencies are generated by a conscious Source.
• Infinity does exist in a pattern at every scale, no matter how large or small.
• The pattern produces unified similarities in projecting the Creation.
• We can be aligned or out of alignment with the pattern.
• Because our consciousness is part of the creative force, and we can self-reflect on our experiences, we can realign with the pattern, harmony, and wholeness.

If I haven't been able to convince you of the existence of other realities, or that Consciousness isn't responsible for the pattern of creation, hopefully I've at least made you curious enough about the possibility that you begin your own investigation. That's great. That's a start.

Now, thanks to science echoing spiritual wisdom, perhaps when you hear things like "God is Light," or "As it is above, so it is below," you can have faith that, if you persevere in your quest, you might just encounter a glimpse of eternal light at the end of the tunnel.

32

In Closing

Religion, Materialism, and The Third Way

How can obscure revelations from the past help us today? How can one practically apply these teachings to their life?

Apperception means taking observations and relating them to past experiences. We process our beliefs according to the weight of past sensory and emotional experiences, which forms how we see the world. An even less efficient filter of reality is adopting the perceptions, ideas, and dogmas of others and taking these beliefs on faith. To this the Gnostics would say "Hold on. Don't take anything on faith until you have tested its truth by your own experience."

And they didn't mean ordinary sensory and emotional experiences. They meant experience gathered from deep meditative states where the senses and emotions take a back seat to transcendental perceptions—states similar to those experienced by the prophets when they had their revelations.

Yes, you can reach such states to greater or lesser degrees, if it's important enough to you. If you seek with your heart and mind, you will eventually have a transformative experience. The problem is, most people have too many distractions and too little desire. It's indicative that, in this age of self-advocacy, the thing we least self-advocate for is our spiritual life.

Our traditional institutions aren't giving us much help either. We're disadvantaged by book-ended orthodoxies—an orthodox version of religion that substituted the liberating mysteries of the ancients with dead-end dogmas and the orthodoxy of scientific materialism that tells us if you can't see it, it doesn't exist.

Most Western societies seem caught between the smug, cynical smog of secular materialism and an equally dulling reliance on old religious dogmas. Neither of these paths provides direct experience or internalization of extraordinary truths. The Third Way is a way out of the dead ends of materialist science and dogmatic religion. It's the unification of science and spiritual wisdom in order to cross-validate new insights and new directions of inquiry. I demonstrate The Third Way in this book by exposing the correlations between spirit, psychology, and science.

This book encourages personal spiritual responsibility. It questions the notion that original sin was incurred by Adam and Eve and passed on to us like a cursed inheritance. The fact is that each of us creates and perpetuates our own "sin" (the actual meaning of which is "error"), and that error is simply ignorance, a forgetting of or disbelief in the original higher state from which the human soul descended.

This book also questions the notion that we must restrict our search for solving existential mysteries with materialistic parameters that, like original sin, limit the more expansive mindsets and tools that allow us to think outside the box. If we were not so fixated on this world of imperfection, we would open our eyes to that state of bliss known as heaven, or as science might term it, profound solutions. That is our error. We chose to be here in this state of misperception; we must choose to work our way out by developing more accurate perceptions. In effect, we must see life through new spiritual and scientific eyes.

Ancient wisdom tells us that we all have divine potential within us, since we are all projections of the Divine Unity (Source). We are all capable of changing, of healing, of helping others, and all these things are miracles. Spiritual presence is like a 24/7 radio frequency, but to find it you have to spend time tuning the dial—we must make that time. Unless we dedicate a portion of each day to study, prayer, meditation, and contemplation, we will keep stumbling along in ignorance.

Spiritual responsibility requires effort, but the rewards are great. At some point in our meditative practice, we build up a critical mass of spiritual energy. Guidance then comes in the form of information gained through dreams, feelings, hunches, and synchronous experiences. New directions become apparent, and miracles start to happen.

MIRACLES COME FROM WITHIN

What are miracles? Miracles don't often happen accompanied by fireworks or glows of light. They are usually more subtle. Miracles are a change of seemingly locked-in patterns. Alcoholics or dope addicts kick their habit. The depressed start to cope without drugs, a long-term illness goes into remission, war and conflict take a detour toward peace—all these events are miracles where habitual patterns that are negative and seem apparent give way to new patterns that are positive, though they may be seemingly unlikely.

The breadth of intellectual rigor and spiritual insights of religions like Judaism and Christianity is immense. But, often, religion has ignored or misapplied core spiritual insights in a kind of cognitive disconnect between their founding revelations and their practice.

If all priests, rabbis, and imams concentrated on and understood their religion's traditions of inner light, the world would be a far better place. Just as we descended from a higher order and contain a spark of higher consciousness, the world's religions descended from a higher common spiritual core. Just as our personal task is to uncover higher consciousness within, the task of religion should be to uncover our spiritual roots.

PROBLEMS WITH MATERIALISM

If anyone thinks materialist science and academia don't share common traits with the religions they've displaced in the modern world, they need to look again.

Materialists have their own brand of orthodoxy, and they protect their turf using a variety of tactics. In academia, they've established a self-constructed standard of academic and political correctness. This squelches debate at the outset because anyone advocating, teaching, or discussing any form of intelligence behind the Creation is stigmatized. Many materialists cloak themselves with the invincibility of "progressive" views, so that even questioning their views is akin to a secular version of heresy.

The term *creationism* means finding ways to support the biblical version of Creation, forcing facts or information to support that viewpoint. Intelligence in creation has nothing to do with creationism. Seeing intelligence behind the Creation is about following the evidence and making a reasonable interpretation of the role of consciousness in the marvelously complex phenomenon known as life.

The Third Way represents a framework or model of reality that is unique and compelling for one simple reason: it is balanced. It uses the subjective and the objective, the rational and the intuitive, the heart and the mind, the left brain and the right brain. It combines science and spirituality. It takes physics and psychology and juxtaposes them on an X and Y axis with science and mysticism to plot out potential answers to life's deepest mysteries.

And life *is* a mystery. No one has all the answers, but The Third Way breaks the one-sided monologues of materialism and religion to synthesize a view supported by both prophets and scientists.

ENLIGHTENMENT IS A PROCESS

Those who are open to alternate perspectives will take what they need from this book for their own personal evolution. Organized religion and materialistic science have added to the human equation both positively and negatively. The negative aspect is that science and religion have packaged their doctrines, beliefs, and traditions. They encourage people to adhere to their party line with the certainty of righteousness on their side. This can stifle growth and inquiry.

Reversing this condition is a *process*. Spiritual, scientific, and intellectual growth is a journey. We need to keep traveling without punching our ticket at anyone else's terminal. A reasonable and intelligent person could look at what has been presented here and come to different conclusions. We can't assume anyone has the absolute truth, just an unfolding series of perspectives on the truth that might be altered or superseded as we grow.

Though we've talked a lot about science, the real thrust of this book is aimed at having a profound spiritual experience. If anything, I hope this book encourages people to take charge of their own spiritual life. Don't blindly believe anyone else (this author included), but respectfully adopt ideas provisionally until you prove or disprove them by your own spiritual experience.

Remember though, your answers will come in symbols, dreams, and actual events in your life, so be alert. There will be no easy-to-identify commands, like your grade-school principal telling you what to do next over the PA system. Sorry. We've got to put away childish things, as the Bible says.

The Gnostics showed we can gain insight equal to or even superior to the discoveries of modern science by subjective inner experience. This is not an article of faith, as in traditional religion. This is tapping into the intelligent energy that created the universe. This is a direct experience with the source of higher information. Knowledge derived by such experience supersedes faith and becomes your own.

True freedom and liberation come from such knowledge and experience. If this book can propel you along that path, it was truly worth me writing and truly worth you reading.

Afterword

This book poses a Third Way, a path between the restrictions of religious dogma and materialist science. Ironically, the evolution from within these disciplines themselves makes this possible, one from recovered knowledge from our past, the other with new discoveries in our present.

However, the current materialist mindset of the world affects us and needs to be addressed. The meaninglessness born from the Neo-Darwinist views that Aldous Huxley described was largely birthed and articulated by intellectuals and academics. Academically, it enshrined Darwinism in the universities. Socially, it spawned philosophies like nihilism and the existentialism of intellectuals like Sartre and Camus. Politically, it supported ideologies like anarchism, socialism, and communism. Today, it has married itself to the idea of secular progressivism.

What's wrong with that? In theory nothing—in practice a lot. These ideologies at their core look entirely at man-made material solutions to problems while ignoring the holistic operations of spirit and consciousness. The problem is what lies at the end of the road of these ideologies. People who drive these agendas become extreme and fanatical about them, and drag a lot of well-meaning people along in their wake. The institutions these people create mostly become dictatorial and repressive, as the driving forces behind them feel the need to strong-arm their beliefs onto society at large.

The promise of institutions taking care of able-bodied people is directly proportional to decreased individual freedoms. This is because highly centralized big government is necessary to take over more aspects of people's lives, which means taking more of their money to fund it, which means huge bureaucracies to govern it. So, bureaucrats and politicians become the new elite in place of the "greedy capitalists." Socialism and communism have been ruinous everywhere they've been tried, in terms of human lives and economics.

Nihilism and existentialism have left the human soul blowing in the winds of random chance. It's tricky because some of these ideologies and philosophies rebelled against things like the stifling repression of orthodox religions. The problem is, they replaced this with things even more harmful and insidious—insidious because they were done under the banner of fraternity, justice, and equality.

We need to ask ourselves whether progressivism in countries like America has eased or made worse immigration crises, economic contractions, and racial divisions. The great Rev. Dr. Martin Luther King Jr. died for a vision of a *color-blind* society. He had a spiritual sense of the concept of unity, but supposedly progressive agendas have pushed civil rights in the opposite direction. Identity politics promoting ideologies such as reparations, affirmative action, critical race theory, and the necessity to tear down the history of a country viewed as inherently evil has divided, not united America. It has mostly benefited politicians and radical groups seeking a power base at the expense of national unity or at least common ground.

Racism is an attitude held in individual hearts and minds. Seeing and defining people through the lens of race is racist. Why? Because if you see every problem, every gripe, every issue you have with someone or some group in racial terms, you are differentiating yourself from them as fellow humans. On campuses we are seeing growing expressions of reverse racism and segregation on the part of people of color. The desire for separate graduation ceremonies, "white free" areas on campuses, and guilt trips fostered by white privilege accusations would dishearten Dr. King or Nelson Mandela, both of whom sought healing in *unity*. These beliefs and attitudes are reactionary mirror images of the old white supremacist legacy.

The issues most Americans view today as racial are really economic at their core. This is why socialism keeps cropping up, despite its failures. The problem with socialism is that its core ideology is a zero-sum game. It presupposes that you need to bring others down to better your lot. No government on Earth owes its people a living. What they owe their people is opportunity and the freedom to choose how they will live. The most positive solution would be to provide everyone with equal opportunity and access to tools for economic growth, not creating dependencies on bloated government.

Emphasis on education and teaching the basics of wealth creation, the establishment of opportunity zones in depressed areas, and business internships for disadvantaged youth are all public-private initiatives that cut across racial lines and don't create government dependencies. Such actions also counter the false promises of socialism and communism that systematize dependencies, destroy initiative, and diminish freedom at the hands of a ruling class of elite political bureaucrats.

We can't heal racial divisions until we stop looking at one another as stereotypes, representatives of past actions, or products of systems, and just look at individuals with an ethos of helping one another—but helping in a way that promotes independence, not dependence.

There are certainly social issues and histories that must be addressed, but we need to address them together, not on opposite sides of the table. You can't build a color-blind society by separating yourself from others on racially based ideologies that underscore differences and hatreds as opposed to working toward common goals. Dr. King fought for equality, not differentiation promoted by guilt, victimization, and negation of people's ethnic or racial identities.

Slavery and genocide of native populations were very real evils that haunt America. Unfortunately, nearly every society on the planet has engaged in these practices—slavery was practiced among Africans and Native Americans long before contact with Europeans. It didn't make all people of those societies bad. The good among them fought these things. Over 300,000 Union soldiers died fighting against slavery.

That's not the hallmark of a systemically bad society, it's the struggle of an **imperfect** society toward a higher good, and without higher consciousness, it can't happen. Higher consciousness encourages working toward unity, but not unity forced by government or fueled by racial division. It requires awakening the presence of a higher force in each of our lives that will guide us on how to live and be of service to our fellow humans.

If societies are to be judged only by the bad things they've done, all of us would stand hopeless and condemned for eternity. That's the mentality of the judgmental Old Testament God in action. If we're to put into practice the spiritual and scientific lessons that we've learned, we need to approach these issues from a different perspective. Each of us is responsible for the state of our own consciousness. If we let ourselves succumb to racially dividing ideologies, we must look ourselves in the mirror after witnessing the results.

Materialist advocates desperately want to keep debate out of the classrooms. The new quantum science and understandings of ancient spiritual wisdom have the potential to expose their soft underbelly, the fact that materiality cannot explain the presence of consciousness let alone explain the phenomena of the quantum world.

So, in academia, materialists have established a self-constructed standard of academic and scientific "correctness." This squelches debate at the outset because anyone advocating, teaching, or discussing intelligence behind the Creation is stigmatized. You are either labeled a Creationist, Believer, Religious Nut, Neanderthal, Primitive, Reactionary, Fanatic or, if you're lucky, a pseudo-scientist for proposing intelligence behind Creation. People have actually lost jobs for this, as mentioned in Ben Stein's documentary *Expelled: No Intelligence Allowed*.

This elitist attitude to impose closure of discussion has people cloaking themselves with the invincibility of logical, progressive views so that even questioning these views is akin to a secular version of heresy. This tyranny derives more from a secular worldview than from hard science. It must change. At the very least, it must be challenged and debated on a fair playing field.

The Third Way is about a unification of science and spirituality that examines life's deepest questions as partners rather than antagonists. It's about rejecting the dogmas of orthodox religion and orthodox science in favor of open inquiry. It's about the ability to debate and seek truth without penalty or stigma.

The Western world, and the world at large, went from the existential frying pan into the fire. The abuses of religion became the perfect target, the scapegoat for a science gone to the other extreme. God is dead, life is meaningless, and the human story is just a random occurrence of particles that somehow formed a peculiar state we call life. Religion threatened people with the fire and brimstone of hell; science threatens people with the soul-sucking vision of thinking, feeling beings trapped in a random, senseless world of suffering and death.

Any rational being has to push back and declare, *this is not acceptable.* And it's certainly not acceptable for our institutions of learning to be churning out future generations programmed with a philosophical and world view that has wide-reaching and detrimental social implications.

This is a ruinous path we should not tread. Religion destroyed many lives, yet science has the potential to destroy far more. Religion didn't invent weapons of mass destruction. Religion didn't invent the humanistic materialist philosophies that led to social systems such as fascism and communism.

In order to create a new paradigm, the first order of business is to stop mixing the concept of conscious intelligence with the biblical, supernatural, Judeo-Christian god. This is the new heresy produced by materialism, the red herring materialists have used to recruit the world to their outlook by playing on people's emotional reaction to religious abuses. Religion once used the same tactic of branding people as heretics in order to destroy them. Gnosticism was destroyed by religion because the God of Gnosticism was *a descent of consciousness,* not the old guy on a throne casting lightning bolts. Science employs more humane punishments, but punishments nonetheless, for bucking the party line.

The next things to acknowledge are the limitations of materialistic science and the scientific method, which cannot grasp what happens beyond space-time from within space-time; neither can it explain the origin of life and consciousness in plausible terms. It cannot even explain the numerous paradoxes present at the quantum level, like traveling faster than the speed of light.

Science has another big flaw—it looks at the world as if humans are separate from the objects we study, rather than being part of a holistic unity with those objects. This limitation is like playing monopoly but being restricted to only rolling the dice every other turn. You're going to miss half the game and a lot of information with such restrictive ground rules. This omission forces you to play the game with one hand tied behind your back, and it yields false conclusions.

Science has no idea what occurred on the other side of the Big Bang singularity, but it stands to reason that the Force that caused the creation of the universe has footprints, echoes, and spillovers into this dimension. We need to expand our vision and methodologies to study and explain not just the world around us but the world *inside* us. Albert Einstein was a great example of a person unafraid to use all his faculties to solve problems. He often intuited possible solutions to questions, then worked backward toward the proofs.

The obvious answer to what lies beyond space-time is consciousness. Accepting the provisional premise that consciousness creates everything and that we are all part of that consciousness gives us a powerful tool to penetrate the true nature of reality and understand our purpose for existing. We can't just "science" our way through tough problems using ordinary left-brained consciousness. Meaningful clues will come from aligning with our non-ordinary higher consciousness through meditation or other practices, tools readily usable for all.

We must recognize that disciplined subjective experience is just as valid, if not more so, than objective scientific approaches in defining the underlying reality of our existential experience. I've demonstrated how the Gnostic masters intuited many theories of quantum physics over two thousand years ago. The point is that cross-referencing objective and subjective information is far better than using either alone in a one-dimensional vacuum.

This merged approach to higher consciousness produces other benefits as well. It puts a measure of control into our seemingly random lives, for once we learn the principles of how higher intelligence works, we begin to focus on those things that have value to us. We even develop a higher notion of what to truly value. We can find new hope in new understandings, making our lives work better by breaking up old, negative patterns. We're incentivized to become more spiritually responsible and aware because we understand that the payoff is putting the power of the universe on our side.

Using science to enhance our spiritual understanding, and spirituality to further our scientific explorations, we can replace the more regressive moral values of religion with a higher spiritual concept of unity and the wholeness within. We can then realize it's in our own self-interest not to hurt others because we see how that ultimately hurts ourselves. Sensing the Source or higher power within us dissolves the deadening void of materialist "meaninglessness" and replaces it with enlightened purpose.

One of the most life-altering experiences is having direct communication with higher intelligence, and receiving information that proves out in the "real" world. Some people become medical intuitives, others develop the intuition that helps their financial or life decisions, some begin a life of service, still others simply do their regular jobs better. The applications are infinite.

When we experience the operation of this information in our lives as it translates from the psychic into the material realm, we have passed beyond *faith* in the unseen (religion) and graduated to *knowledge* that the unseen is the true underlying reality (spirituality).

Few people understand the real meaning of quantum physics, and fewer still understand the hidden spiritual wisdom of the past. If these concepts go mainstream, they will transform our world and exponentially propel the evolution of our consciousness. If you doubt this statement, consider this—a few hundred years ago, the church had its grip around the neck of the truth. Science gradually bled through this iron curtain. In our lifetime, we saw the accelerated erosion of such millennia-old fear-based concepts as the belief in a judgmental biblical God.

Traditional science, for better or worse, has shaped the basic world outlook of most people today. But science replaced the old concept of God with an empty shell of chance occurrences—life, the Great Accident, which is not such a great life after all. We went from a rock to a hard place.

But a new genie is now out of the bottle, and it's taking the form of quantum physics echoed by rediscovered ancient gospels shedding light on a universal wisdom tradition. It provides a truer picture of creation, reality, and our human origin and purpose. It's an empowering message that transcends the limitations of traditional religion and the dulling atheism of materialistic science. It will take time to filter into the mass mind's awareness, but when it catches on, we'll see a quantum leap in all aspects of human activity as we discard the yokes of our past religious and scientific beliefs.

Charles Sanders Peirce, a pioneer of the modern scientific empirical method, wisely said that the conclusions of science are always tentative, and the scientific method should be self-corrective. With this mindset and continued application of this method, science can detect and correct its own mistakes, and thus eventually lead to the discovery of truth. Many scientists have forgotten this wisdom. They've fallen into a religious-like orthodoxy of static, dogmatic beliefs that stifle alternative views.

What a great mystery and a great relief it would be if we could realize that the "God" behind quantum physics might well be the God of the ancient mystics come to reclaim a place in our hearts and our awareness. It might dispel the ignorant errors of our previous works, and that is true evolution in action, the continual correction and gradual gaining of insights into the meaning of our origins. Though we may or may not achieve the final perspective on these matters, perhaps the closest we will come to discovering universal truth is the point where the objective world of science intersects with the subjective world of mystic spirituality.

The last thought I will leave you with is to put the work in, then *let go and don't overthink* issues. Miracles happen spontaneously, even unconsciously, in between our thoughts. Trust in the process. Thoughtlessness is needed at some point, to let the information flow. Physicist Amit Goswami said:

> The evidence for God is within us, but to see it we have to be subtle . . . if you will write without thinking of the result in terms of a result, but think of the writing in terms of discovery . . . creation must take place between the pen and the paper, not before in a thought or afterwards in a recasting.[1]

Discovery is what life is about. Learn, focus, and experiment in your contemplative moments, then relax. The universe is infinite intelligence. It doesn't need obsessive-compulsive control freaks to worry things along.

I'd wish you good fortune, but you'll now begin to create your own with no further help needed. All I can wish for you is to START NOW.

Notes

INTRODUCTION

1. Luke 17:21.

CHAPTER 1. GETTING STARTED

1. Gospel of Thomas, saying 70, Nag Hammadi Library, Gnostic Society Library website, n.d.

CHAPTER 3. THE MATERIALIST WORLDVIEW

1. Richard Lewontin, "Billions and Billions of Demons," New York Review of Books website, January 9, 1997.
2. George Wald, "The Origin of Life," *Scientific American* website, August 1954, 44–53.
3. Richard Dawkins, *The Selfish Gene,* 2nd ed. (Oxford: Oxford University Press, 1976), 1.
4. Stephen Jay Gould, "Evolution's Erratic Pace," *Natural History* 86, no. 5 (May 1977): 12–16.
5. Stephen L. Talbott, "Evolution: A Third Way?" Nature Institute website, 2015.
6. Vernon Blackmore and Andrew Page, *Evolution, the Great Debate* (London: Lion, 1989), 79, 82.
7. John van Wyhe, ed., "*Origin of Species* Variorum," Darwin Online, 2002.
8. Gerald. L. Schroeder, *The Science of God: The Convergence of Scientific and Biblical Wisdom* (New York: Broadway Books, 1998), 38.
9. Schroeder, *The Science of God,* 37.
10. James Hull and Gerhard Wagner, eds. *Aldous Huxley, Representative Man* (Münster, Germany: Lit Verlag, 2004), 321.

CHAPTER 4. THE PERENNIAL PHILOSOPHY

1. Francis Legge, ed., *The Refutation of All Heresies,* vol. 1 (London: Society for Promoting Christian Knowledge, 1921), chap. 20.

CHAPTER 5. THE SACRED FEMININE

1. Jean Markale, *The Great Goddess: Reverence of the Divine Feminine from the Paleolithic to the Present* (Rochester, Vt.: Inner Traditions, 1999), 51.

CHAPTER 7. THE GREAT GNOSTIC MYTH OF SOPHIA

1. J. J. Hurtak, "Pistis Sophia Text with Commentary," Pistis Sophia: Text and Commentary website. Accessed April 21, 2022.

CHAPTER 8. THE FALL

1. G. R. S. Mead, trans., "Pistis Sophia," division 1, book 1, chapters 32–42, Gnostic Society Library website, 15–19.
2. Hans-Gebhard Bethge and Bentley Layton, trans., "On the Origin of the World," Nag Hammadi Library, Gnostic Society Library website, n.d.
3. Frederik Wisse, trans., "The Apocryphon of John," Nag Hammadi Library, Gnostic Society Library website, n.d.

CHAPTER 9. THE FORCE OF OPPOSITION

1. Wesley W. Isenberg, trans., "The Gospel of Philip," Nag Hammadi Library, Gnostic Society Library website, n.d.
2. Hans-Gebhard Bethge and Bentley Layton, trans., "On the Origin of the World," Nag Hammadi Library, Gnostic Society Library website, n.d.

CHAPTER 10. THE TRUTH OF THE MATTER

1. Philip Schaff, ed., "Fragments of Papias," in *The Apostolic Fathers with Justin Martyr and Irenaeus,* vol. 1 of Ante-Nicene Fathers, Christian Classics Ethereal Library website, n.d.

CHAPTER 11. JEWISH MYSTICISM

1. Rabbi Nissan Dovid Dubov, "Moses," Kabbalah, Chassidism and Jewish Mysticism, at Chabad website, n.d.
2. E. H. Gifford, trans., "Eusebius of Caesarea: Præparatio Evangelica," book 8, Tertullian website, November 2003.
3. Hippolytus, *Refutatio Omnium Hæresium,* ed. Miroslav Marcovich (Berlin: De Gruyter, 2014), 18–28.
4. Thomas Taylor, trans., *Iamblichus' Life of Pythagoras* (Rochester, Vt.: Inner Traditions International, 1986), 8.

5. Taylor, trans., *Iamblichus' Life of Pythagoras,* 8.

6. Fred Gladstone Bratton, *A History of the Bible: An Introduction to the Historical Method* (Boston: Beacon Press, 1967), 79–80.

7. Hugh Schonfield, *The Passover Plot* (New York: Disinformation Co., 2005), chap. 2.

8. Flavius Josephus, *The Jewish War,* Book 2, 8.11.154 (London: Penguin, 1988).

9. Josephus, *The Jewish War,* Book 2, 8.12.159.

10. Josephus, *The Jewish War,* Book 2, 8.7.137.

11. Wesley W. Isenberg, trans., "The Gospel of Philip," Nag Hammadi Library, Gnostic Society Library website.

12. Mark 2:23.

CHAPTER 12. GNOSTIC JESUS?

1. Mark 4:11.

2. Morton Smith, *Clement of Alexandria and a Secret Gospel of Mark* (Cambridge, Mass: Harvard University Press, 1973), 446.

3. Smith, *Clement of Alexandria,* 446.

4. Origen, "Contra Celsus" 9:18, Nag Hammadi Library, Gnostic Society Library website, n.d.

5. Bhagavad Gita, chapter 4, verses 8–9, at Srimad Bhagavad Gita website, n.d.

CHAPTER 15. WHAT WE LOST

1. Stevan Davies, trans., Gospel of Thomas, verse 114, Nag Hammadi Library, Gnostic Society Library website, n.d.

2. Stephen Emmel, trans., "Dialogue of the Savior," Nag Hammadi Library, Gnostic Society Library website, n.d.

3. Marvin Meyer, trans., Gospel of Philip, Nag Hammadi Library, Gnostic Society Library website, n.d.

4. G. R. S. Mead, trans., "Pistis Sophia," book 1, chap. 17, Gnostic Society Library website, n.d.

5. Mead, trans., "Pistis Sophia," book 1, chap. 34.

6. Jacques Berlioz, *Tuez-les tous, Dieu reconnaîtra les siens: Le massacre de Béziers (22 juillet 1209) et la croisade contre les Albigeois vus par Césaire de Heisterbach.* Loubatières, 1994.

CHAPTER 16. JOURNEY OF THE MIND VOYAGER

1. Gerhard Adler, William McGuire, and Herbert Read, eds., *The Collected Works of C. G. Jung,* trans. R. F. C. Hull (Princeton, N.J.: Princeton University Press. 2014), 384.

2. Lyn Cowan, ed., *Barcelona 2004: Edges of Experience: Memory and Emergence; Proceedings of the 16th International Congress for Analytical Psychology* (Einsiedeln: Daimon, 2006), 915.

3. "Introduction," Nag Hammadi Library, Gnostic Society Library website. n.d.

4. C. G. Jung and Joseph Campbell, *The Portable Jung* (New York: Viking Press, 1971), 407–8.

5. Gospel of Thomas, saying 22, Nag Hammadi Library, Gnostic Society Library website, n.d.

6. Carl Jung, "Inner Voice," lecture presented at Kulturbund, Vienna, 1932.

7. *Collected Works of C. G. Jung,* vol. 7, *Two Essays in Analytical Psychology,* trans. Gerhard Adler and R. F. C. Hull, 2nd ed., (Princeton, N.J.: Princeton University Press, 2014), 68.

8. Gospel of Thomas, saying 70, Nag Hammadi Library, Gnostic Society Library website.

9. *Collected Works of C. G. Jung,* vol. 9 (Part 2), *Aion: Researches into the Phenomenology of the Self,* trans. Gerhard Adler and R. F. C. Hull, ed. William McGuire, Sir Herbert Read, Michael Fordham, (Princeton, N.J.: Princeton University Press, 2014), para. 126.

CHAPTER 17. FROM MIND TO MATTER

1. *Collected Works of C. G. Jung,* vol. 8, *Structure & Dynamics of the Psyche,* trans. Gerhard Adler and R. F. C. Hull, 2nd ed., (Princeton, N.J.: Princeton University Press, 2014) 382–84.

2. Jung, *Structure & Dynamics of the Psyche,* 747.

3. C. G. Jung, *Memories, Dreams, Reflections,* ed. Aniela Jaffe, trans. Clara Winston and Richard Winston, (New York: Vintage Books, 1989), 345.

4. C. G. Jung, "Commentary on 'The Secret of the Golden Flower,'" in *Alchemical Studies,* vol. 13 of *Collected Works of C. G. Jung* (Princeton, N.J.: Princeton University Press, 1983), 45.

CHAPTER 19. BEHOLD THE QUANTUM

1. Max Planck, "The Nature of Matter," lecture given at Florence, Italy, 1944.

CHAPTER 20. I AM, THEREFORE I THINK

1. Gregg Braden, "DNA Experiments by Gregg Braden," Fulvio Frisone: The Official Website, n.d.

2. Amit Goswami, "The Science of Quantum Consciousness," lecture at Schumacher College, England, 2012.

CHAPTER 22. TO BE OR NOT TO BE

1. Emerging Technology from the arXiv, "A Quantum Experiment Suggests There's No Such Thing as Objective Reality," MIT Technology Review website, April 2, 2020.
2. The Circuit Mojo HD, "Michio Kaku: Time Travel, Parallel Universes, and Reality," YouTube website, May 25, 2008.

CHAPTER 23. THROUGH A GLASS, DARKLY

1. 1 Corinthians 13:12, King James Version.

CHAPTER 24. WORLDS OF POSSIBILITIES

1. Marina Jones, "David Bohm and the Holographic Universe," Futurism website, March 31, 2014.

CHAPTER 25. SWIMMING IN ENERGY

1. Walter Thirring, "On Physics (1960–1969)," Quotable Mathematics website, January 22, 2021.
2. University of Virginia, "Are We on the Brink of a New Age of Scientific Discovery?" SciTechDaily website, May 10, 2021.
3. D. Kazanis, "The Physical Basis for Subtle Bodies and the Near-Death Experience," *Journal of Near-Death Studies* 14, no. 2 (Winter 1995): 101–16.

CHAPTER 26. CONSCIOUSNESS, THE HARD PROBLEM

1. James Herrick, *The Making of the New Spirituality: The Eclipse of the Western Religious Tradition* (Downers Grove, Ill.: InterVarsity Press, 2003), 172–73.
2. Max Planck, "The Nature of Matter," lecture given at Florence, Italy, 1944.
3. James Jeans, *The Mysterious Universe* (Cambridge University Press, 1930), 137.
4. Arthur Stanley Eddington, "Reality," chap. 13 in *The Nature of the Physical World* (Cambridge Univ. Press, 1927).
5. Kevin Dickinson, "Did We Evolve to See Reality as It Exists? No, Says Cognitive Psychologist Donald Hoffman," Big Think website, August 9, 2019.
6. David Bohm, "Essay," in *Wholeness and the Implicate Order* (London: Routledge, 2008), 14.
7. John Wheeler and Martin Redfern, "The Anthropic Universe," interview, ABC Radio National website, Australian Broadcasting Corporation, February 22, 2006.
8. Literary Hub, "Philip Goff and Philip Pullman Talk Materialism, Panpsychism, and Philosophical Zombies," Literary Hub website, November 4, 2019.

9. Tam Hunt, "Could Consciousness All Come down to the Way Things Vibrate?" ed. Beth Daley, The Conversation website, November 9, 2018.

10. Albert Einstein and Leopold Infeld, *Evolution of Physics: The Growth of Ideas from the Early Concepts of Relativity and Quanta* (Cambridge: Cambridge University Press, 1938), 33.

CHAPTER 27. THREE THOUSAND YEARS LATE

1. G. R. S. Mead, trans., "Pistis Sophia," book 2, chapter 68.4, Gnostic Society Library website.

2. Mead, trans., "Pistis Sophia," book 2, chapter 69.9.

3. Mead, trans., "Pistis Sophia," book 1, chapter 32.14.

4. Mead, trans. "Pistis Sophia," book 1, chapter 39.

5. Hans-Gebhard Bethge and Bentley Layton, trans., "On the Origin of the World," Nag Hammadi Library, Gnostic Society Library website, n.d.

6. Bethge and Layton, trans., "On the Origin of the World."

7. Frederik Wisse, "The Apocryphon of John," Nag Hammadi Library, Gnostic Society Library website, n.d.

CHAPTER 28. NEVER MIND

1. Frederik Wisse, trans., "The Apocryphon of John," Nag Hammadi Library, Gnostic Society Library website, n.d.

2. Jan Erik Sigdell, "Is Yahweh an Anunnaku?" Epilogue, Christian Reincarnation website, September 25, 2010.

3. "The First Book of IEOU," Gnostic Scriptures and Fragments, Gnostic Society Library website, n.d.

CHAPTER 29. WHAT ABOUT ME?

1. Bob Samples, *The Metaphoric Mind: A Celebration of Creative Consciousness* (San Francisco, Calif.: Addison-Wesley Publishing, 1976), 26.

AFTERWORD

1. Amit Goswami, "God Is Not Dead Quotes by Amit Goswami," Goodreads website, 2022.

Index